Automate Everything With Python

Mastering Python for Real-World Automation: Transform Workflows, Simplify Tasks, and Boost Productivity.

Automate Everything With Python
Mastering Python for Real-World Automation: Transform Workflows, Simplify Tasks, and Boost Productivity

For permission requests, write to the publisher at:
Jeff Miller
https://jeffmillerauthor.net/

Table of Contents

Chapter 0:
Setting Up Your Environments

Welcome, aspiring automation enthusiasts! You're about to embark on an exciting journey into the world of Python programming and data automation. This book is not your typical Python introduction. We won't waste time diving into the history or evolution of Python or slog through exercises that hold no relevance in the professional world. Instead, every page is focused on practical, real-world skills that will transform how you work.

Here, you'll learn how to create efficient, automated processes that solve real business challenges. We'll explore how to:

- Execute SQL queries and Python scripts seamlessly.
- Download content from the web.
- Upload and download data to and from SQL databases.
- Work with text, csv and Excel files.
- Create and format Excel reports.
- Generate monthly workflows.
- Email reports as attachments.
- Automate the download of files from Outlook emails.
- Generate professional-grade PDF files.
- Automate a backup solution to protect critical files.
- Schedule your code to run automatically, so your workflows can operate hands-free.

This book is designed for professionals who want to bring automation into their daily tasks. The focus is on how to develop solutions that actually work in a professional environment. By the end of this journey, you'll have the skills to build processes that can dramatically reduce manual effort and improve efficiency.

For example, using these same principles in my work, I've taken a process that used to take 8 hours to complete and reduced it to 58 seconds. I've also transformed a laborious, three-month process into one that now takes only 15 minutes. These are the kinds of practical, impactful results you can achieve.

This book is specifically designed for Windows users, walking you step-by-step through setting up your development environment and building powerful automation solutions. Most chapters are designed to stand on their own, meaning you won't need to reread the entire book if you need to reference how to accomplish a specific task. Each chapter clearly identifies the libraries and tools required for the given task, ensuring you can pick up right where you need. Let's get started!

How to Use This Book

Throughout this book, you'll encounter two special icons:

 (Little Bee Blasting Off): This icon indicates a section that you can skip if you're already familiar with the content. It's perfect for more experienced users who want to move quickly through the material.

 (Little Bee with Stop Sign): When you see this icon, it's a signal to slow down and pay close attention. These sections contain crucial information that you should read carefully, regardless of your experience level.

If you're already comfortable with Python and have a preferred setup, feel free to look for the icons and skip to the next section with the Little Bee with Stop Sign. For everyone else, let's get started with installing Anaconda and exploring Spyder, our main development tool.

Preparing Your Database Environment

Before we dive into automating with Python, we need to ensure we have a database ready for our exercises. We'll be working with a database named 'Automate_Everything_With_Python'. While it's possible to create a database using Python, we'll create this database manually using SQL Server Management Studio (SSMS). This approach ensures that you can manually set up the database, especially since we're not familiar with your permissions and role in the system. It also allows you to become more familiar with SQL Server Management Studio (SSMS), a crucial tool for database management, and helps you understand the structure of SQL Server databases visually.

Creating a Database in SQL Server Management Studio (SSMS)

Follow these steps to manually create a database named 'Automate_Everything_With_Python' in SSMS:

1. **Open SQL Server Management Studio (SSMS)**
   - Launch SSMS from your Start menu or desktop shortcut.
2. **Connect to Your SQL Server Instance**
   - In the "Connect to Server" dialog, select your server name.
   - Choose "Windows Authentication" or enter your SQL Server credentials if using SQL Server Authentication.
   - Click "Connect".
3. **Create the New Database**
   - In the Object Explorer (usually on the left), right-click on "Databases".
   - Select "New Database" from the context menu.

4. **Set Database Properties**
   - In the "New Database" dialog, enter "Automate_Everything_With_Python" as the database name.
   - Leave other settings at their default values for now.
   - Click "OK" to create the database.

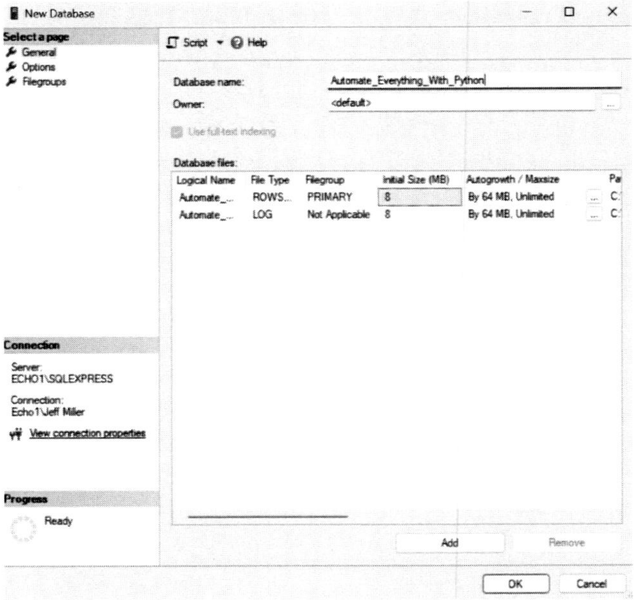

5. **Verify Database Creation**
   - In the Object Explorer, refresh the "Databases" folder if necessary.
   - You should see "Automate_Everything_With_Python" listed under "Databases".

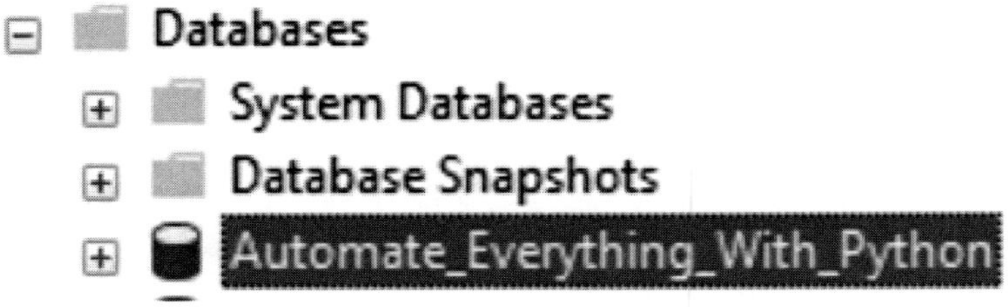

Congratulations! You have now created a new database named 'Automate_Everything_With_Python' in SQL Server Management Studio.

By setting up the database manually, you're creating a controlled environment that our Python scripts will interact with. We'll use Python to query data from this database and insert data into it. This separation of database setup and Python scripting mirrors real-world scenarios where you often work with existing databases. Now that we have our database ready, let's move on to setting up a user for the database.

Creating Your Database User

Now that we have our 'Automate_Everything_With_Python' database set up, it's time to create a dedicated user for our Python scripts. In the world of database management, it's best practice to use specific users for specific purposes, rather than always relying on admin accounts.

For our exercises, we'll create a user named 'automation_user'. This user will have admin rights to our database, allowing our Python scripts to perform all necessary operations. In a real-world scenario, you'd typically use more restricted permissions, but for learning purposes, we'll keep things simple.

Follow the steps below to set up this user in SQL Server Management Studio (SSMS). Pay close attention to each step, as proper user set-up is crucial for the security and functionality of your database operations.

Setting Up a User in SQL Server Management Studio (SSMS)

Follow these step-by-step instructions to create a new user named 'automation_user' with admin rights to the 'Automate_Everything_With_Python' database:

1. Create a New Login

- In the Object Explorer (usually on the left), expand your server node.
- Right-click on the "Security" folder.
- Select "New" > "Login".

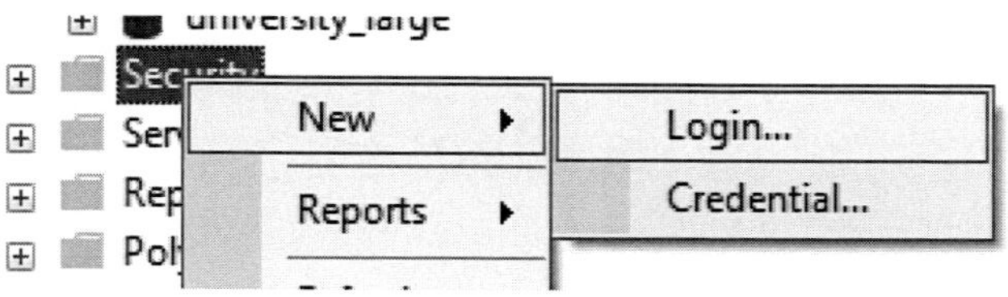

2. Set Up the Login Properties

- In the "Login - New" dialog:
- For "Login name", enter: automation_user
- Select "SQL Server authentication"
- Enter the password: AutomateEverything2024!
- Uncheck "Enforce password policy" (for this exercise only; in real-world scenarios, always use strong password policies)
- Set "Default database" to Automate_Everything_With_Python

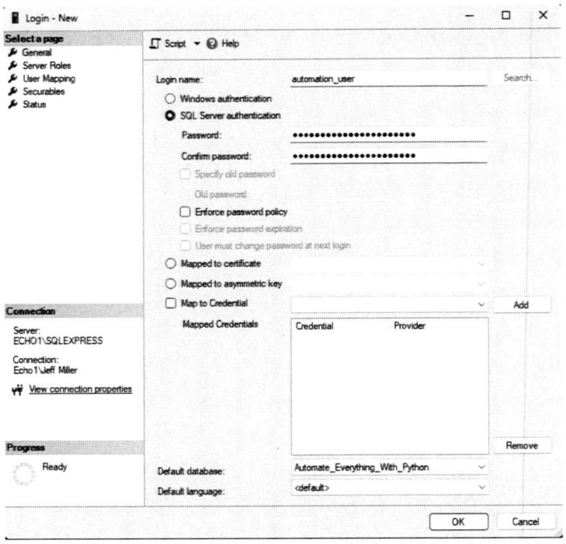

3. Grant Server Roles (Optional, for full admin rights)

- In the same "Login - New" dialog, go to the "Server Roles" page.
- Check "sysadmin" to grant server admin rights. (Note: This grants extensive permissions. In a production environment, you'd typically use more restricted roles.)

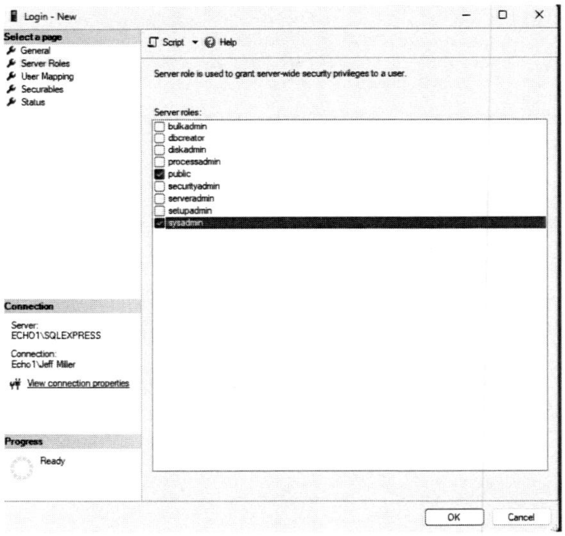

4. Set Up User Mapping

- Go to the "User Mapping" page.
- Check the box next to Automate_Everything_With_Python
- Enter automation_user for "User"
- In the "Database role membership" section for this database, check: db_owner (This gives full rights to the database)

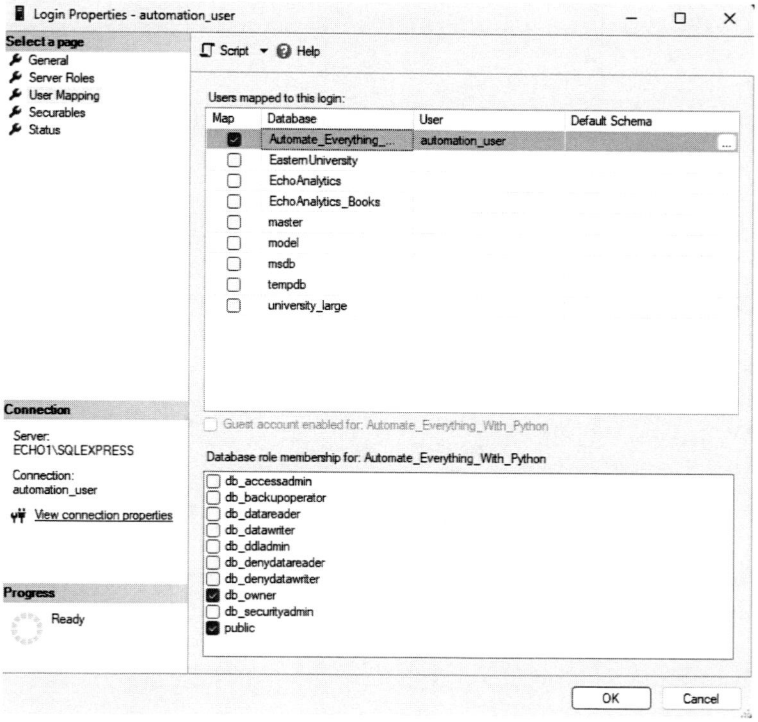

5. Finalize the User Creation

- Click "OK" to create the login and user.

6. Verify the User Creation

- Expand the "Security" > "Logins" folder in Object Explorer.
- You should see automation_user listed.

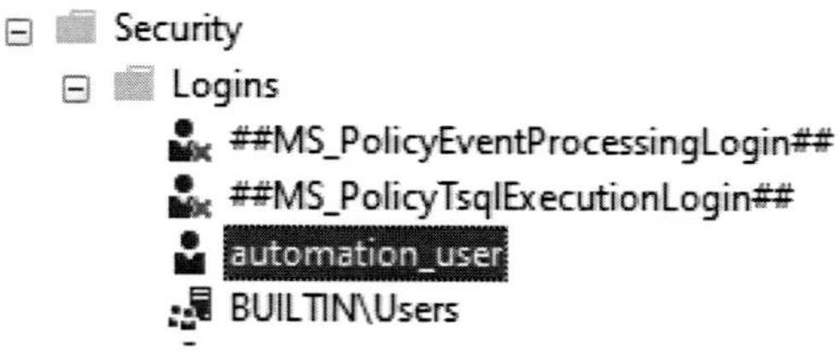

7. Test the New Login

- Disconnect from the server (File > Disconnect)
- Connect again, but this time use:
  - Authentication: "SQL Server Authentication"
  - Login: automation_user
  - Password: AutomateEverything2024
- If you can connect and see the Automate_Everything_With_Python database, the setup was successful.

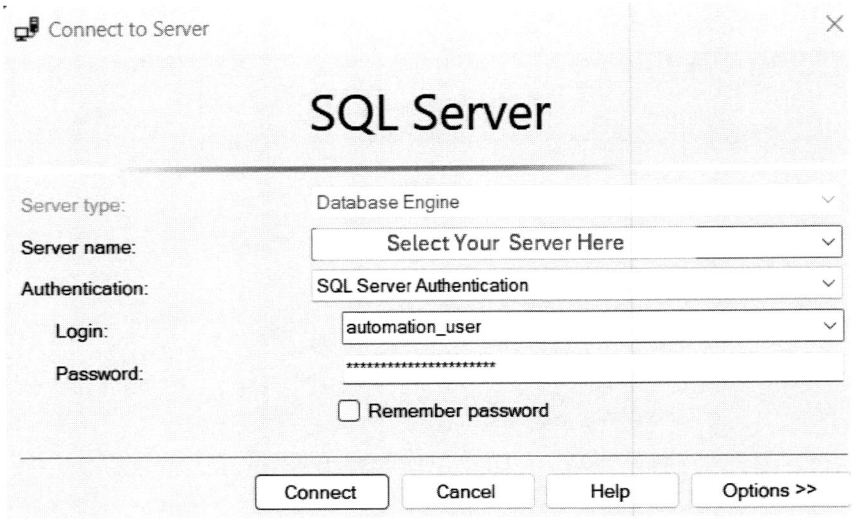

By following these steps, you've created a user that your Python scripts can use to interact with the database throughout the exercises in this book.

With these steps completed, you've not only created a database but also a user with the right permissions to interact with it. This setup will be the foundation for all the exciting Python automation tasks we'll tackle in the coming chapters.

Remember, the power to create and manage users comes with great responsibility. Always prioritize security in your database management practices!

Why Choose Anaconda?

Anaconda offers several benefits for both beginners and experienced users:

- It includes Python, so you don't need a separate installation.
- It comes with pre-installed libraries commonly used in data science and scientific computing.
- It includes Spyder, an Integrated Development Environment (IDE) that we'll use throughout this book.
- It provides an easy-to-use package and environment management system.

Installation Steps

Follow these steps to install Anaconda and launch Spyder:

Step 1: Download Anaconda

1. Open your web browser and go to the Anaconda website: https://www.anaconda.com/products/distribution
2. After the page loads, you'll see a registration form. You have two options:
   a. If you'd like to receive updates and information from Anaconda, you can fill out the form and click "Submit".
   b. If you prefer not to register, simply click the "Skip registration" link located directly under the "Submit" button
3. Click on the "Download" button, which will appear after either submitting the form or skipping registration.
4. Select the appropriate version for your operating system.
5. Choose the latest Python 3.x version (ignore any Python 2.x versions if present).
6. Your download will begin automatically. If it doesn't start, look for a direct download link on the page.

Step 2: Install Anaconda

1. Find the downloaded installer file (e.g., "Anaconda3-202X.XX-Windows-x86_64.exe"). It will be in your Downloads folder.
2. Double-click the file to start the installation.
3. Follow the prompts, accepting the license agreement.
4. Choose "Just Me" for the installation type unless you want to install for all users.
5. Select an installation location or use the default.
6. In "Advanced Options," check both:

   a. "Create shortcuts (supported packaged only)"
   b. "Register Anaconda3 as my default Python"

7. Click "Install" and wait for the process to complete.
8. Click "Next" twice, then "Finish."
9. Anaconda will launch shortly after.

Step 3: Launch Spyder

1. Click the Start menu, find Anaconda Navigator, and launch it. Then click on the Spyder icon.

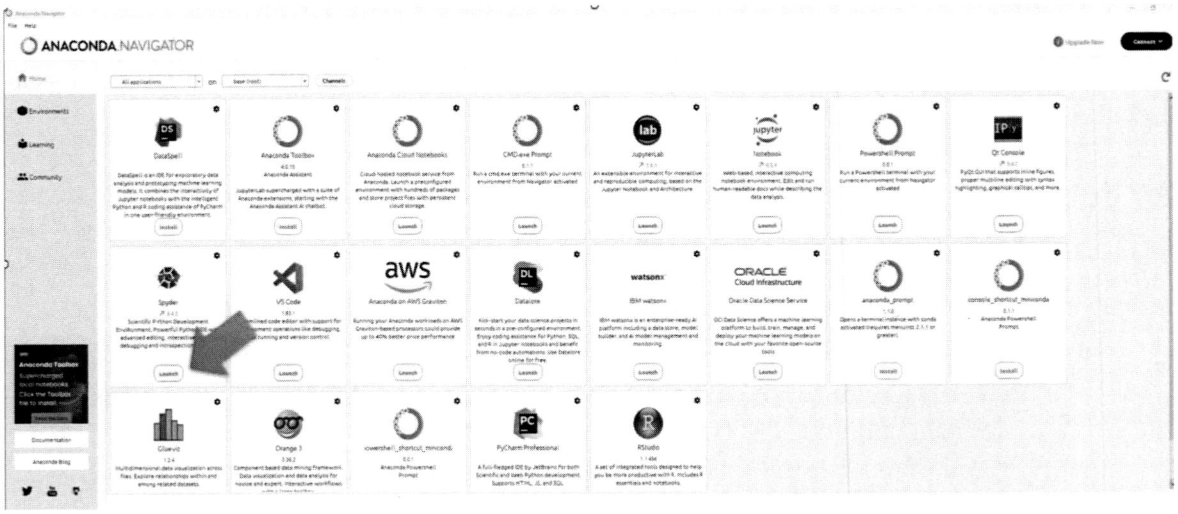

2. Alternatively, search for "Spyder" in your computer's applications and launch it directly.
3. When you open Anaconda, you may see a window to sign in or sign up. You can skip this if you prefer.

Getting Started with Spyder

Now that we have Spyder open, let's explore its main features and write our first Python code.

Understanding Spyder's Layout

Spyder has several important sections:

1. **Editor**: This is where you'll write your Python code.
2. **Console**: Here you'll see the results of your code and can run quick, one-line commands.
3. **Variable Explorer**: This shows you the contents of your variables.
4. **File Explorer**: Use this to navigate your computer's files.
5. **Run File Icon (Green Play Button)**: When you click this button, Spyder will execute all the code in the current file, and the output will be displayed in the Console at the bottom of the screen. This button is best used when you want to run the complete script from start to finish.
6. **Run Current Line or Selection Icon (Green Arrow with a 'Step In' Icon)**: This option is helpful for quickly testing small portions of code without running the entire script. You can select any part of the code and use this green arrow to execute only that selection.
7. **Interrupt Kernal (Gray/Red Square Icon)**: This button provides a great visual indicator of your code's execution status. When your code is running, the button turns red, signaling that the Python interpreter is actively processing your script. If your code

completes naturally, the button will turn gray again, indicating that execution has finished. This color change from red to gray is an excellent visual cue, letting you know the current state of your code execution. It's particularly useful for managing long-running scripts or identifying infinite loops. If your code is taking too long or seems stuck, you can click this red square button to interrupt the kernel and stop the code from running. When you click it, you're essentially sending an interruption request to the Python interpreter, forcing it to terminate the current execution.

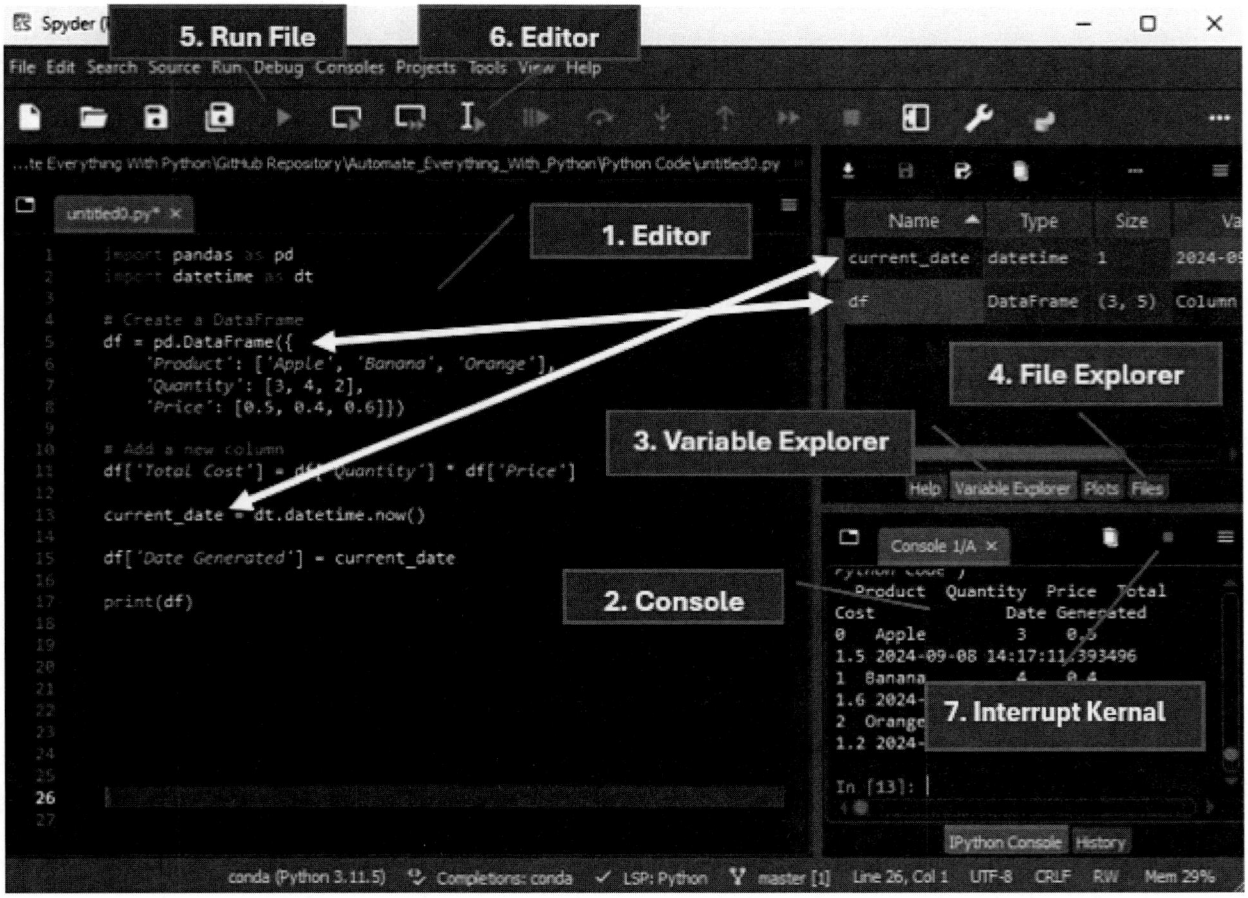

Python Library Installation Guide for Anaconda Spyder

This section will guide you through setting up all the necessary libraries for the entire book. We'll check if the required libraries are already installed and provide instructions on how to install any that are missing using Anaconda Spyder. Remember, once a library is installed, you won't need to install it again for future chapters.

Checking Installed Libraries

1. Open Anaconda Spyder.
2. Create a New File
   - Click File, New File…
3. In the IPython console, type and run the following code to check which libraries are already installed:

```
# Test what libraries are installed
import pandas

libraries = ["pandas", "pyodbc","sqlalchemy","openpyxl","dateutil","win32com"]


for lib in libraries:
    try:
        __import__(lib)
        print(f"{lib} is installed")
    except ImportError:
        print(f"{lib} is NOT installed")
```

This code will print messages in the IPython Console indicating whether each library is installed or not.

Library Descriptions

Here's a brief description of each library we'll be using:

1. **pandas**: A powerful data manipulation and analysis library. It provides data structures like DataFrames for efficiently handling large datasets.
2. **pyodbc**: A module that makes accessing ODBC databases simple. It implements the DB API 2.0 specification.
3. **sqlalchemy**: A SQL toolkit and Object-Relational Mapping (ORM) library for Python. It provides a full suite of well-known enterprise-level persistence patterns.
4. **openpyxl**: A library for reading and writing Excel 2010 xlsx/xlsm/xltx/xltm files.
5. **dateutil**: A powerful extension to the standard datetime module, available in the python-dateutil library.
6. **win32com**: Part of the pywin32 extensions, it provides access to many of Windows' COM (Component Object Model) interfaces.

Installing Libraries

If any libraries are not installed, follow these steps:

1. Open Anaconda Prompt (you can search for it in the Windows start menu).
2. Here are the specific installation commands for each library:

```
conda install pandas
conda install pyodbc
conda install sqlalchemy
conda install openpyxl
conda install python-dateutil
pip install pywin32
```

3. After installation, restart Anaconda Spyder to ensure all changes take effect.

Your First Python Code: Creating Comments, Basics of DataFrames and Working with Dates

The code example will demonstrate how to create comments, work with a **Pandas DataFrame**, retrieve the current date, and interact with Spyder's **Variable Explorer** to better understand how variables are stored and processed.

A Glimpse of Your Goal

Before we dive into the details, let's take a peek at what we're aiming to create. The following code snippet shows the complete Python code you'll have written by the end of this chapter:

```python
# Importing Required Libraries
import pandas as pd
from datetime import datetime

# Creating a dictionary of data
data = {'Name': ['John', 'Anna', 'Peter', 'Linda'],
        'Age': [28, 24, 35, 32],
        'City': ['New York', 'Paris', 'Berlin', 'London']}

# Creating a DataFrame
df = pd.DataFrame(data)

# Retrieving the current date
current_date = datetime.now().date()

# Adding a new column to the DataFrame with the current date
df['Date Retrieved'] = current_date

# Print DataFrame
print(df)
```

This code snippet demonstrates how to import libraries, create a DataFrame, add a date column, and view the results. Don't worry if it looks a bit complex right now - we'll break it down step-by-step and explain each part in detail.

As you progress through this chapter, you'll gain the skills to write and understand this code, setting a solid foundation for your Python journey. Let's begin by breaking down each component.

Comments and Libraries

First, let's start by importing the necessary libraries in the **Editor**. You'll notice that the code below includes comments (lines starting with #). While you don't have to type these comments, it's good practice to include them in your code for several reasons:

1. They help you and others understand what the code is doing.
2. When time passes, it's easy to forget the purpose of specific code sections, and comments serve as reminders.
3. Comments aid debugging by providing context for each section of your code.
4. They offer searchable context. If you're trying to find a snippet of code that you can't quite remember, you can use Ctrl+F to search for a relevant term in your comments.

```python
# Importing Required Libraries
import pandas as pd
from datetime import datetime
```

Explanation:

- **Pandas:** is a Python library that provides data structures and functions needed to work with structured data (such as data in tables or spreadsheets). Its most common data structure is the DataFrame, which is like an Excel spreadsheet or a SQL table.
- **Datetime:** allows us to work with dates and times.
- **#:** The # Importing Required Libraries is a comment that provides a description of what the following lines of code do. Comments are marked with a # symbol in Python, meaning that the interpreter ignores them when executing the code. Here, the comment indicates that this section is about importing the libraries needed to work with data (pandas) and handle dates (datetime).
    - To add a comment, simply type # before your text. This is useful for explaining parts of your code to yourself or others.
    - In **Spyder**, you can add or remove comments by selecting the row(s) and pressing **Ctrl + 1** to toggle comments.

Creating a Sample DataFrame from a Dictionary

Let's break down the process of creating a DataFrame from a dictionary:

```
# Creating a dictionary of data
data = {'Name': ['John', 'Anna', 'Peter', 'Linda'],
        'Age': [28, 24, 35, 32],
        'City': ['New York', 'Paris', 'Berlin', 'London']}

# Creating a DataFrame
df = pd.DataFrame(data)
```

This code involves two main steps: creating a dictionary and converting it to a DataFrame.

Creating a Dictionary

Here's what's happening:

- We create a variable called data.
- The {} curly braces indicate that we're creating a dictionary.
- In Python, a dictionary is a collection of key-value pairs.
- Each key in the dictionary (e.g., 'Name', 'Age', 'City') represents a column name in our future DataFrame.
- The values associated with each key are lists containing the data for each column.

Let's break it down further:

- The 'Name' key has a value that is a list of strings: ['John', 'Anna', 'Peter', 'Linda']
- The 'Age' key has a value that is a list of integers: [28, 24, 35, 32]
- The 'City' key has a value that is another list of strings: ['New York', 'Paris', 'Berlin', 'London']

Each list has the same number of elements (4 in this case), which will correspond to the number of rows in our DataFrame.

Converting the Dictionary to a DataFrame

```
# Creating a DataFrame
df = pd.DataFrame(data)
```

This line does the magic of converting our dictionary into a DataFrame:

- We use the pd.DataFrame() function, which is part of the pandas library imported as pd.
- We pass our data dictionary as an argument to this function.
- The resulting DataFrame is assigned to the variable df.

The pd.DataFrame() function interprets the dictionary in the following way:

- Each key in the dictionary becomes a column name.
- Each value (which is a list) becomes the data for that column.
- The position of each item in the list determines which row it belongs to.

After this operation, df will be a DataFrame with:

- 3 columns: 'Name', 'Age', and 'City'
- 4 rows of data, each containing a name, age, and city

You can visualize it like this:

Name	Age	City
John	28	New York
Anna	24	Paris
Peter	35	Berlin
Linda	32	London

This method of creating a DataFrame is particularly useful when you have data that's already organized into categories (like names, ages, and cities in this case). It allows you to quickly create a structured, tabular representation of your data that's easy to manipulate and analyze using pandas' powerful functions.

Adding Current Date to DataFrame

Let's break down the process of retrieving the current date and adding it to our DataFrame:

```
# Retrieving the current date
current_date = datetime.now().date()

# Adding a new column to the DataFrame with the current date
df['Date Retrieved'] = current_date
```

This code involves two main steps: creating a variable with the current date and then using this variable to add a new column to our DataFrame.

Retrieving the Current Date

Here's what's happening in this line:

```
# Retrieving the current date
current_date = datetime.now().date()
```

- **datetime.now():** This function call returns a datetime object representing the current date and time.
  - datetime is a class from the datetime module we imported earlier.
  - now() is a method that returns the current local date and time.
- **.date():** This method extracts just the date part from the datetime object.
  - It removes the time information, giving us only the year, month, and day.
- **current_date =:** We assign the resulting date to a variable named current_date.

After this operation, current_date will contain today's date. For example, if you run this code on September 15, 2023, current_date would contain the date 2023-09-15.

Adding a New Column to the DataFrame

```
# Adding a new column to the DataFrame with the current date
df['Date Retrieved'] = current_date
```

This line adds a new column to our DataFrame. Let's break it down:

- **df:** This is our DataFrame that we created earlier.
- **['Date Retrieved']:** This syntax is used to create a new column in the DataFrame.
  - If 'Date Retrieved' doesn't exist as a column, this will create it.
  - If it does exist, this will overwrite its contents.
- **current_date:** We assign the value of current_date to this new column.

The magic of pandas is that it automatically extends this single date value to all rows in the DataFrame. So if your DataFrame had 100 rows, all 100 rows in the new 'Date Retrieved' column would contain the same date.

After this operation, your DataFrame df will have a new column called 'Date Retrieved', and every row in this column will contain the current date.

This method of adding a column is particularly useful when you want to timestamp your data or add any constant value across all rows of your DataFrame. It's a common operation in data preprocessing and can be very helpful in tracking when data was collected or processed.

Viewing Your DataFrame

After creating and manipulating your DataFrame, you'll often want to inspect its contents. In Spyder, there are two primary ways to view your DataFrame: using the print() function and utilizing the Variable Explorer. Let's explore both methods:

Method 1: Using the print() Function

The most straightforward way to view your DataFrame is by using the print() function in your code. Here's how you do it:

```
# Print DataFrame
print(df)
```

When you run this line, the contents of your DataFrame will be displayed in the console output. It will look like this:

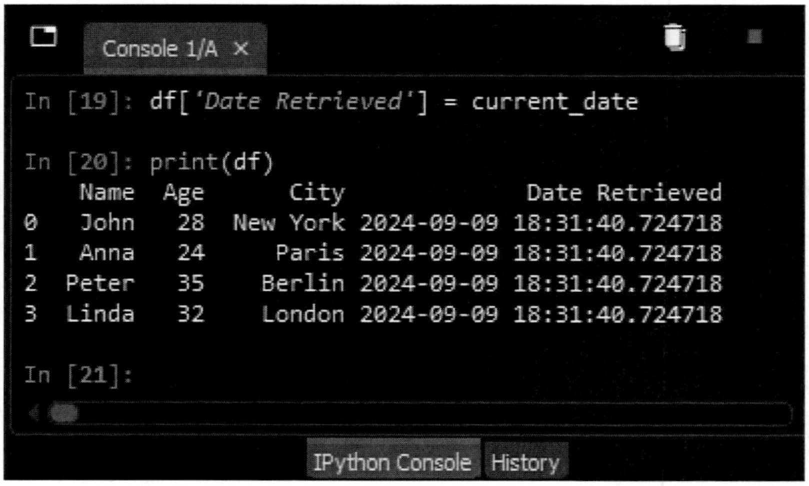

Method 2: Using Spyder's Variable Explorer

My favorite, the Variable Explorer in Spyder, is a powerful tool for inspecting your data. After running the code above:

1. Look at the Variable Explorer pane.
2. You'll see variables like df and current_date listed.
3. Double-click on df to open a spreadsheet-like view of your DataFrame.

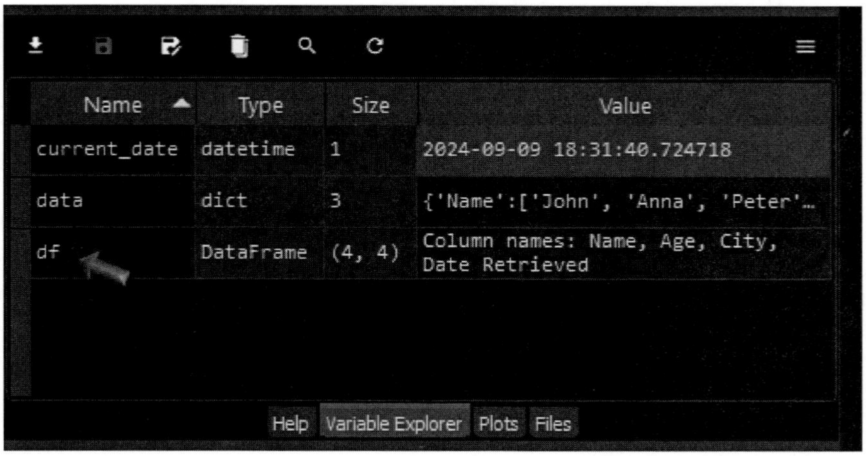

4. You can sort, filter, and inspect the data in this view.

The Variable Explorer allows you to:

- Inspect DataFrames and other complex data structures visually.
- Monitor how variables change as your code runs.
- Interact with your data, making development and debugging more intuitive.
- Sort DataFrames by clicking the header or column name.

Conclusion

Congratulations! You've taken your first steps into the world of Python programming with Anaconda and Spyder. Let's recap what you've accomplished:

- Successfully installed Anaconda, providing you with a robust Python environment.
- Launched Spyder, a powerful Integrated Development Environment (IDE) for Python.
- Learned about Spyder's layout and key features, including the Editor, Console, and Variable Explorer.
- Written your first Python code, including:
    - Creating comments for better code readability
    - Importing essential libraries like pandas and datetime
    - Creating a DataFrame from a dictionary
    - Working with dates and adding them to your DataFrame
    - Viewing your data using both the print() function and Spyder's Variable Explorer

These foundational skills are crucial as you begin your journey in Python programming and data analysis. The Variable Explorer will be a valuable tool as you work with more complex data structures and debug your code.

As you move forward, remember to:

- Experiment with the code examples provided
- Use comments to explain your code
- Leverage Spyder's features, especially the Variable Explorer, to understand your data better
- Practice creating and manipulating DataFrames with different types of data

In the next chapter, we'll build on these fundamentals.

Chapter 1:
Extracting Files from GitHub

Introduction to Extracting Files from GitHub

Welcome back! In this chapter, we'll automate the process of cloning a GitHub repository to your local machine using Python. This skill is invaluable for developers and data scientists who frequently need to download and work with code from various sources. Instead of manually downloading files from GitHub, we'll automate the process—saving time and effort.

By the end of this chapter, you will know how to clone a repository programmatically using Python, ensuring you always have the latest version of the data and code for your projects. This automation fits perfectly into our broader goal: **Automate Everything!**

Before we begin, if you haven't installed Git, we'll briefly walk through the installation process. Having Git installed is essential for this automation to work. Let's dive in!

What is Git?

Git is a distributed version control system that allows multiple people to collaborate on projects, especially software development. It tracks changes in the code, enabling developers to revert to previous versions if needed, create separate branches for feature development, and collaborate seamlessly across teams.

Key Features of Git:

- **Version Control**: Git keeps track of all changes made to a codebase. You can go back to previous versions if needed or compare changes.
- **Branching**: Developers can create new branches to work on features or bug fixes without affecting the main codebase. This ensures that parallel work can happen safely.
- **Collaboration**: Multiple people can work on the same project simultaneously without overwriting each other's work.
- **Distributed**: Every developer has a full copy of the repository, including the entire history of changes, ensuring that work can continue even when not connected to a central server.

Git is a powerful tool in the software development world because it simplifies collaboration and helps maintain the integrity of code. It's widely used in development projects hosted on platforms like GitHub, Bitbucket, and GitLab.

Why Do We Need Git?

In the context of this book, we are using Git to clone repositories from GitHub. This allows us to efficiently download the files we need for our exercises, ensuring that we always have the most up-to-date materials.

Using Git gives us a few specific advantages:

- **Automation**: Instead of manually downloading individual files from GitHub, we can automate the process of pulling entire repositories.
- **Version Control**: We ensure that we're always working with the latest version of the files, and if there are updates, we can easily fetch them using Git's commands.
- **Consistency**: When working with multiple people or environments, Git helps keep everyone on the same page by managing changes across the project.

Without Git, downloading and keeping track of files from large projects can be tedious and error prone. By automating this process, we ensure that our workflow is both efficient and reproducible.

Step 1: Installing Git (If needed)

If you haven't installed Git yet, follow these steps for your operating system.

1. **Download Git**: Visit git-scm.com and download the Windows installer.
2. **Install Git**: Run the installer and accept the default settings.
3. **Verify the installation**:
    a. Open **Command Prompt** and type git --version. You should see the installed version of Git.

With Git installed, we're ready to move on to Python coding.

Code Overview: Automating Repository Cloning

Before diving into the code, let's take a high-level look at the script we'll be creating. This script will:

- Define the GitHub repository to be cloned.
- Set up the paths where the repository will be saved.
- Check if the repository already exists on your machine.
- Clone the repository if it's not already present.

By the end of this chapter, you'll have an automated way to download GitHub repositories and working files straight to your desktop. Let's get started step by step!

```python
import os
import subprocess

# Define the GitHub repository URL
repo_url = "https://github.com/JeffMillerAuthor/Automate_Everything_With_Python"

# Set the desktop path and destination path
desktop_path = os.path.join(os.path.expanduser("~"), "Desktop")
repo_name = repo_url.split("/")[-1]  # Keep the original repository name
destination_path = os.path.join(desktop_path, repo_name)

# Check if the destination path already exists
if not os.path.exists(destination_path):
    # Clone the repository to the destination path
    subprocess.run(["git", "clone", repo_url, destination_path], check=True)
    print(f"Repository cloned successfully to {destination_path}")
else:
    print(f"The folder '{destination_path}' already exists on your desktop.")
```

Step 2: Setting Up the Python Script

Next, let's create a Python script that automates the process of cloning a GitHub repository.

Importing Libraries

Start by importing the necessary libraries:

```python
import os
import subprocess
```

Explanation

- **os**: This module allows us to interact with the operating system, such as reading and writing to directories.
- **subprocess**: This module lets us run system commands directly from Python, which we will use to execute Git commands.

Summary

- Imported the os and subprocess libraries to interact with the operating system and run Git commands from within Python.

Step 3: Setting Up File Paths and Repository URL

Now, we define the GitHub repository URL and where we want to save the cloned files.

```
# Define the GitHub repository URL
repo_url = "https://github.com/JeffMillerAuthor/Automate_Everything_With_Python"

# Set the desktop path and destination path
desktop_path = os.path.join(os.path.expanduser("~"), "Desktop")
repo_name = repo_url.split("/")[-1]  # Keep the original repository name
destination_path = os.path.join(desktop_path, repo_name)

```

Explanation

- **repo_url**: This is the URL of the GitHub repository you want to clone.
- **os.path.expanduser("~")**: Returns the path to the current user's home directory.
- **os.path.join()**: Joins directory names to form a complete path, ensuring compatibility across operating systems.
- **repo_name**: Extracts the repository name (e.g., Automate_Everything_With_Python) from the URL.
- **destination_path**: Combines the desktop path with the repository name to create the full path where the repository will be cloned.

Summary

- Defined the GitHub repository URL and set up paths to clone the repository to the desktop.

Step 4: Cloning the Repository

Now let's handle the cloning process. The script will check if the folder already exists on your desktop and clone the repository if it's not there.

```
# Check if the destination path already exists
if not os.path.exists(destination_path):
    # Clone the repository to the destination path
    subprocess.run(["git", "clone", repo_url, destination_path], check=True)
    print(f"Repository cloned successfully to {destination_path}")
else:
    print(f"The folder '{destination_path}' already exists on your desktop.")
```

Explanation

- **os.path.exists(destination_path)**: Checks if the folder already exists. If not, the script proceeds with cloning.

- **subprocess.run(["git", "clone", repo_url, destination_path], check=True)**: Runs the git clone command using the subprocess module. The check=True argument ensures that an error is raised if the cloning fails.
- **Print statements**: These provide feedback to the user on whether the repository was cloned or if the folder already exists.

Summary

- The script checks if the repository already exists and clones it if necessary. Clear feedback is provided to the user.

Final Output

Upon running the code, the "Automate Anything with Python" directory will be created on your desktop, containing the following essential files and subfolders from the GitHub repository

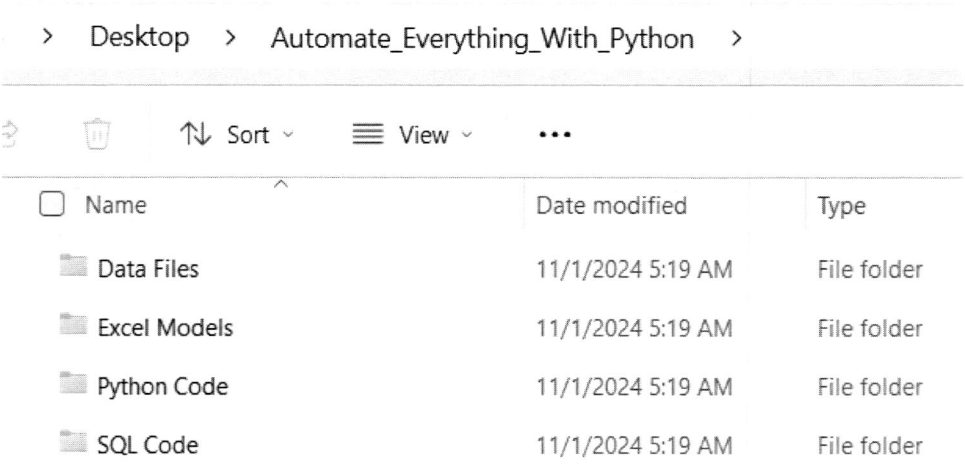

Data Files: This folder holds flat files (text, CSV, and Excel formats) that serve as the input data for SQL loading and email processing. Python will loop through these files, generating and sending emails to designated recipients.

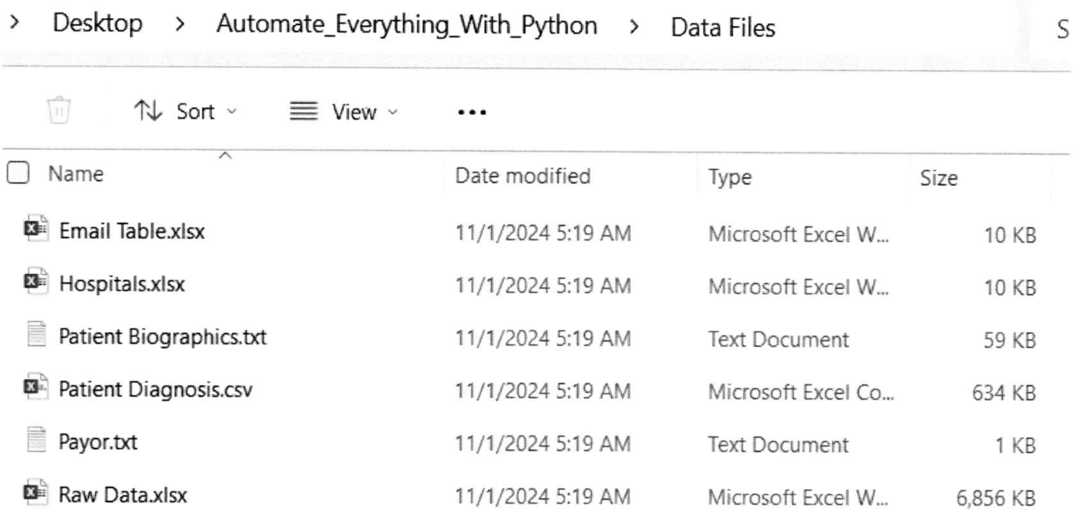

Excel Files: Here, you'll find an Excel workbook with embedded macros. Python will trigger these macros and VBA scripts to automate tasks, such as generating PDF statements.

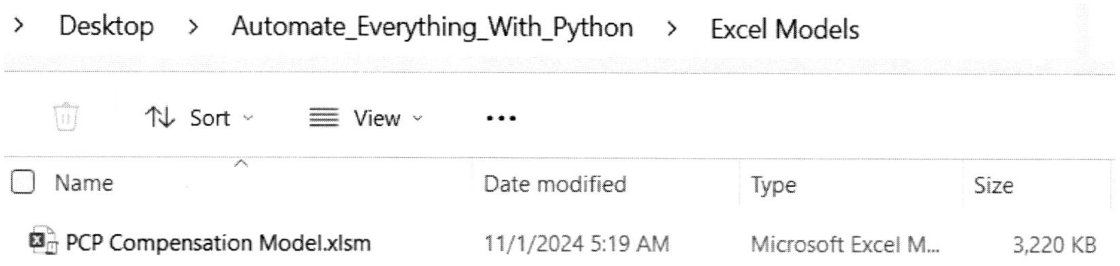

Python Code: This folder contains all the Python scripts covered in each chapter, allowing you to easily reference and apply the code as you progress.

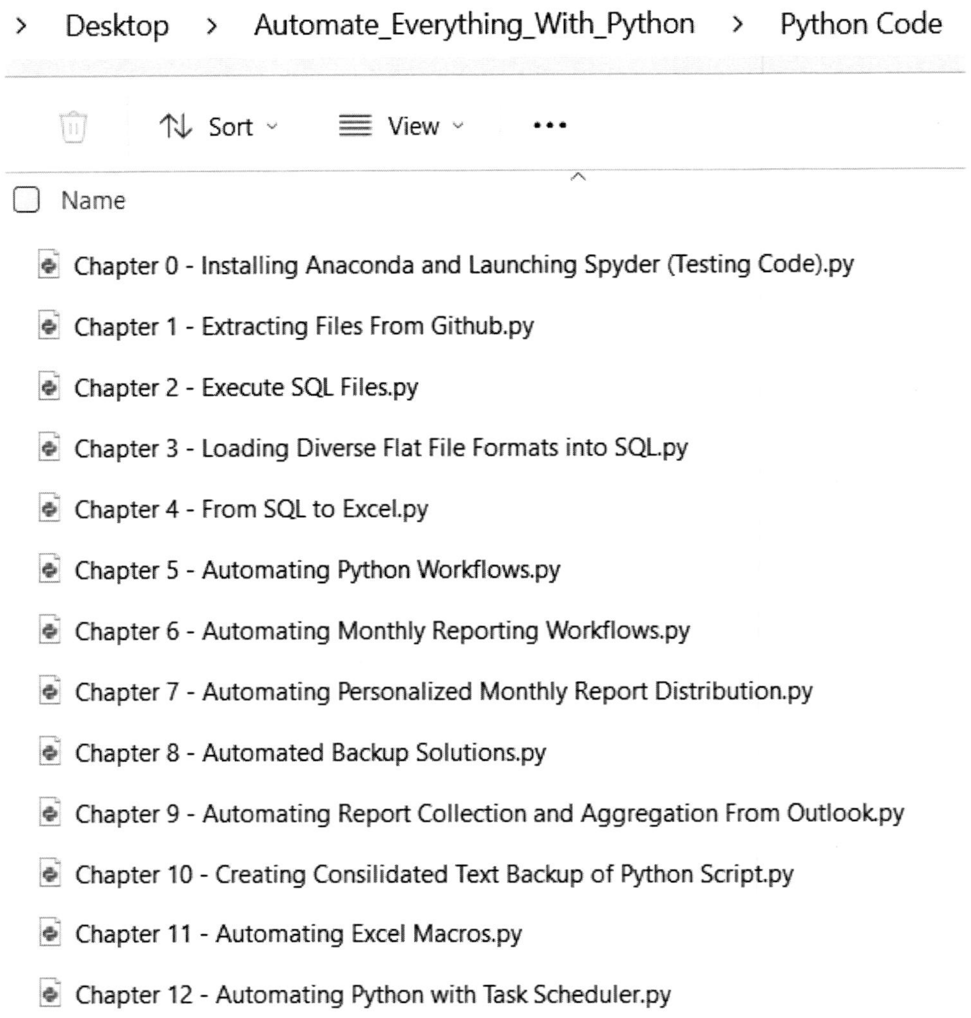

SQL Code: This folder includes SQL scripts that Python will execute to load data into SQL Server. You may also view and edit these files directly in SSMS if desired.

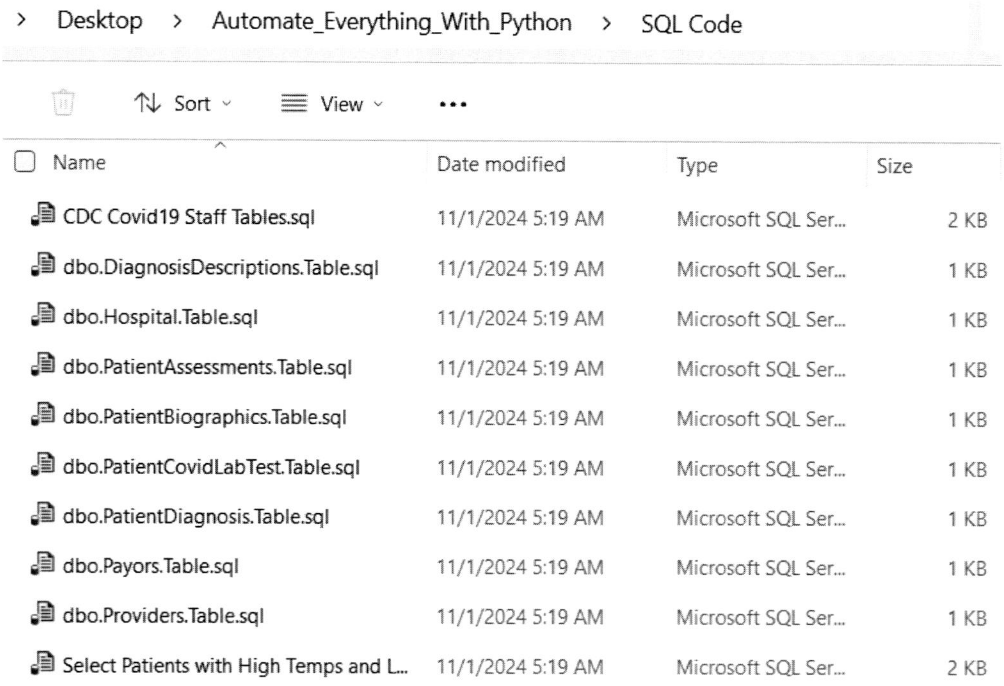

These resources will be essential as we move through the rest of the book, providing hands-on materials for practicing automation techniques.

Conclusion

In this chapter, you've learned how to automate the process of cloning a GitHub repository to your local machine using Python. Here's a quick recap of what we've covered:

- **Installed Git** (if necessary) to allow for programmatic interactions with GitHub.
- **Set up paths and repository details**: Defined the repository URL and set the destination path on the desktop.
- **Cloned the repository**: Automated the process of cloning a repository and provided feedback to the user.

This script ensures that you always have the latest version of your project's code without having to manually download files from GitHub. This sets the stage for more advanced automation techniques, which we will explore in the upcoming chapters.

As we proceed, you'll use the files you've just cloned to dive deeper into Python scripting, data processing, and automation. Get ready for hands-on exercises that will take your skills to the next level!

In the next chapter, we'll start working with the files from this repository, beginning with data extraction and processing techniques in Python. Let's keep the automation journey going!

Chapter 2:
Executing SQL Files

Introduction to SQL File Execution Automation

Welcome back, data automation enthusiasts! In this chapter, we'll dive into automating SQL file execution, creating a powerful Python script that connects to a SQL Server database, reads SQL files from a specified directory, executes them, and processes the results, all within the Anaconda Spyder environment. This automation is especially useful for database administrators, data analysts, and developers who handle multiple SQL scripts and need to execute them efficiently.

By the end of this chapter, you will be able to:

- Connect to a SQL Server database.
- Read SQL files from a directory.
- Execute SQL files programmatically.
- Handle errors and provide feedback on execution.

Let's embark on this journey to supercharge your SQL workflow!

Code Overview: Automating SQL File Execution

Before diving into the details, here's a high-level overview of the steps you'll take:

- Import necessary libraries.
- Set up a connection to SQL Server.
- List SQL files in a directory.
- Execute SQL scripts one by one.
- Handle potential errors gracefully.

The "Chapter 2 - Execute SQL Files.py" script we'll build can be found in the "Python Code" folder in your "Automate_Everything_With_Python" directory. The following visual is a reflection of the code we are about to develop in context. Now, let's break the script down into manageable steps.

```python
import pyodbc
import os

# Define the connection details
server = 'Enter_Your_Server_Name'  # Replace with your SQL Server name
database = 'Automate_Everything_With_Python'  # Database name where SQL files will be executed
sql_username = 'automation_user'  # Replace with your SQL Server username
password = 'AutomateEverything2024!'  # Replace with your SQL Server password

# Get the username
pc_username = os.getlogin()

# Path to the directory containing SQL files
sql_directory = r"C:\\Users\\" + pc_username + "\\Desktop\\Automate_Everything_With_Python\\SQL Code"

# List of SQL file names to execute
sql_files = [f for f in os.listdir(sql_directory) if (f.startswith("dbo") or f.startswith("CDC Covid19 Staff")) and f.endswith(".sql")]

# Define the connection string
connection_string = f'DRIVER={{ODBC Driver 17 for SQL Server}};SERVER={server};DATABASE={database};UID={sql_username};PWD={password}'

# Function to execute SQL file
def execute_sql_file(file_path):
    try:
        # Establish a connection and execute the SQL script
        with pyodbc.connect(connection_string) as conn:
            with open(file_path, 'r') as file:
                sql_script = file.read()
                with conn.cursor() as cursor:
                    cursor.execute(sql_script)
                    print(f"SQL script {os.path.basename(file_path)} executed successfully.")
    except pyodbc.Error as e:
        print(f"Error executing {os.path.basename(file_path)}: {e}")
    except FileNotFoundError:
        print(f"File not found: {file_path}")
    except Exception as e:
        print(f"An error occurred with {os.path.basename(file_path)}: {e}")

# Execute each SQL file
for sql_file in sql_files:
    sql_file_path = os.path.join(sql_directory, sql_file)
    execute_sql_file(sql_file_path)
```

Step 1: Importing Libraries and Setting Up the Environment

We start by importing two essential libraries—os for file handling and pyodbc for database connections.

```python
import pyodbc
import os
```

Explanation

- **os**: A standard Python module that allows interaction with the operating system, including handling file and directory paths, creating, renaming, and deleting files.
- **pyodbc**: A Python library for working with databases using ODBC drivers. It enables connections to databases like SQL Server and executing queries.

Summary

In this step, we imported os to manage file operations and pyodbc to connect to and interact with a SQL Server database.

Step 2: Setting Up the SQL Server Connection and Directory

Next, we define the connection details to connect to the SQL Server database. You will need to enter the server name, which you can find instructions in the Explanation section below.

```
# Define the connection details
server = 'ECHO1\\SQLEXPRESS'  # Replace with your SQL Server name
database = 'Automate_Everything_With_Python'  # Database name where SQL files will be executed
sql_username = 'automation_user'  # Replace with your SQL Server username
password = 'AutomateEverything2024!'  # Replace with your SQL Server password

# Get the username
pc_username = os.getlogin()

# Path to the directory containing SQL files
sql_directory = r"C:\\Users\\" + pc_username + "\\Desktop\\Automate_Everything_With_Python\\SQL Code"
```

Explanation

- **Connection Details**:
  - server: The name of the SQL Server instance.
    - Getting your server name:
      - **Method 1: Connect to Server Window**
        When you first open SQL Server Management Studio (SSMS), the **Connect to Server** window will appear. In this window, locate the **Server name** field. The server name is typically in the format `COMPUTER_NAME\SQLEXPRESS` if you're using a local instance. If the field is blank or incorrect, you can click the dropdown arrow to check for previously connected servers.
    - **Method 2: Object Explorer**
      If you're already logged in and need to check the server name, you can find it in the **Object Explorer** panel on the left side of SSMS. If the panel isn't visible, you can enable it by selecting **View > Object Explorer** in the top menu. The server name will appear at the top of the **Object Explorer**, showing the currently connected server.
  - database: The specific database where SQL scripts will be executed.
  - sql_username and password: Credentials used to log into the SQL Server.
- **os.getlogin()**: Retrieves the username of the currently logged-in user. This is used to construct the file path dynamically, so the script works for any user.

Summary

In this step, we defined the connection details needed to establish a connection with SQL Server, including how to locate your server name and input the correct database, username, and password. Additionally, we used os.getlogin() to dynamically build file paths for SQL scripts. This setup enables Python to connect to the correct database and access the necessary SQL files for execution.

Step 3: Listing and Filtering SQL Files

Now that the connection setup is ready, let's read and filter the SQL files in the target directory.

```
# List of SQL file names to execute
sql_files = [f for f in os.listdir(sql_directory) if (f.startswith("dbo") or f.startswith("CDC Covid19 Staff")) and f.endswith(".sql")]
```

Explanation

- **os.listdir(sql_directory)**: Lists all files and directories in the specified directory path (sql_directory).
- **f for f in ...**: A list comprehension that iterates through each file or directory name (f) returned by os.listdir.
- **f.startswith("dbo") or f.startswith("CDC Covid19 Staff")**: Checks if the current file name starts with either "dbo" or "CDC Covid19 Staff". This ensures only files with these prefixes are considered.
- **f.endswith(".sql")**: Ensures the file name ends with the .sql extension, filtering out non-SQL files.
- **if ...**: A condition within the list comprehension that includes only files meeting the specified criteria.
- **sql_files**: The resulting list that contains all files from sql_directory matching the defined filters.

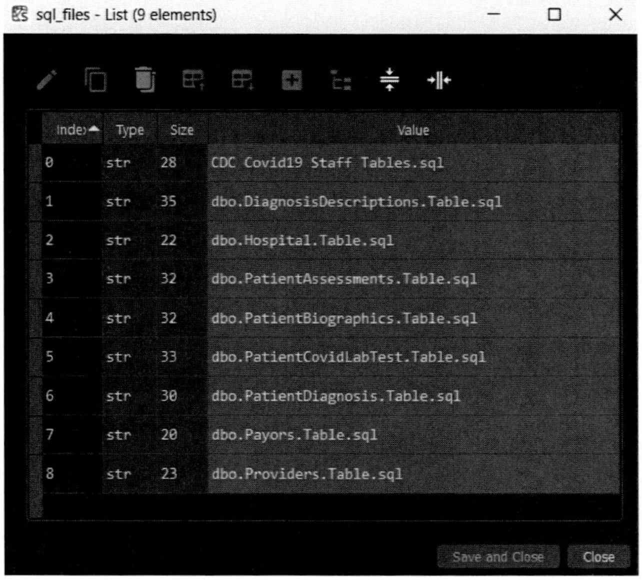

Index	Type	Size	Value
0	str	28	CDC Covid19 Staff Tables.sql
1	str	35	dbo.DiagnosisDescriptions.Table.sql
2	str	22	dbo.Hospital.Table.sql
3	str	32	dbo.PatientAssessments.Table.sql
4	str	32	dbo.PatientBiographics.Table.sql
5	str	33	dbo.PatientCovidLabTest.Table.sql
6	str	30	dbo.PatientDiagnosis.Table.sql
7	str	20	dbo.Payors.Table.sql
8	str	23	dbo.Providers.Table.sql

Summary

In this step, we used os.listdir() to list all SQL files in the directory and applied filtering with list comprehension to select only the files that start with either "dbo" or "CDC Covid19 Staff" and end with ".sql". This ensures that only relevant SQL scripts are included for execution, making the process more efficient and organized.

Step 4: Defining the SQL Connection String

Now that we've filtered the SQL files, let's set up the connection string that will allow Python to communicate with SQL Server.

```
# Define the connection string
connection_string = f'DRIVER={{ODBC Driver 17 for SQL Server}};SERVER={server};DATABASE={database};UID={sql_username};PWD={password}'
```

Explanation

This code sets up a connection string that contains the details necessary for connecting to a SQL Server database using Python. A connection string is essentially a set of parameters that tells Python how to connect to the database.

- **connection_string**: This variable holds all the details needed to establish the connection to the SQL Server.
- **f'...' (f-string)**: This is a format string (f-string) in Python, which allows variables to be inserted into the string. The placeholders {server}, {database}, {sql_username}, and {password} are replaced with the actual values during script execution.
- **DRIVER={{ODBC Driver 17 for SQL Server}}**:
  - **ODBC Driver 17**: This is the driver that enables Python to communicate with SQL Server. It's crucial for establishing the connection between Python and SQL Server.
  - **Double Curly Braces {{...}}**: In Python f-strings, double curly braces are used to escape literal curly braces. This ensures that the ODBC driver syntax is maintained as required by SQL Server.
- **SERVER={server}**: The SERVER parameter specifies the address of the SQL Server that the script will connect to. In the code, the server variable will hold the name or IP address of the server.
- **DATABASE={database}**: This specifies the database you are connecting to, in this case, the Automate_Everything_With_Python database. This allows Python to query, update, and manipulate data within that database.
- **UID={sql_username}**: **UID** stands for User ID, which is the username used for authentication. In this case, the username is automation_user. This user account must have the appropriate permissions to access and execute SQL queries on the database.
- **PWD={password}**: **PWD** stands for Password, which is the password corresponding to the sql_username. This ensures secure access to the database. In this case, the password is 'AutomateEverything2024!'.

Summary

In this step, we created a connection string that contains all the necessary details to connect Python to SQL Server. The connection string includes the driver, server address, database name, user credentials (UID), and password (PWD). This setup allows Python to interact with SQL Server and execute SQL queries.

Step 5: Creating the SQL Execution Function

In this step, we will create a Python function to execute SQL files, handle potential errors, and provide feedback on the execution status.

```
# Function to execute SQL file
def execute_sql_file(file_path):
    try:
        # Establish a connection and execute the SQL script
        with pyodbc.connect(connection_string) as conn:
            with open(file_path, 'r') as file:
                sql_script = file.read()
                with conn.cursor() as cursor:
                    cursor.execute(sql_script)
                    print(f"SQL script {os.path.basename(file_path)} executed successfully.")
    except pyodbc.Error as e:
        print(f"Error executing {os.path.basename(file_path)}: {e}")
    except FileNotFoundError:
        print(f"File not found: {file_path}")
    except Exception as e:
        print(f"An error occurred with {os.path.basename(file_path)}: {e}")
```

Explanation

Purpose: This Python function, execute_sql_file, is designed to:

- Open a given SQL file.
- Establish a connection to a SQL Server database.
- Execute the SQL script contained in the file.
- Handle any errors that may occur during the process.

This function is particularly useful when automating the execution of multiple SQL files as part of a larger automation process.

```
def execute_sql_file(file_path):
    try:
        # Establish a connection and execute the SQL script
        with pyodbc.connect(connection_string) as conn:
            with open(file_path, 'r') as file:
                sql_script = file.read()
                with conn.cursor() as cursor:
                    cursor.execute(sql_script)
                    print(f"SQL script {os.path.basename(file_path)} executed successfully.")
```

Explanation

- **Function Definition**:
  - This line defines a function called execute_sql_file. The function takes one parameter, file_path, which is the path to the SQL file you want to execute.
- **Try-Except Block**:
  - The entire function is wrapped in a try-except block to handle errors gracefully.

- **Why?** This ensures that if something goes wrong (e.g., the file is not found or there's a SQL execution error), the program won't crash. Instead, it will handle the error and display an informative message.

- **Establishing a Database Connection**:
  - **pyodbc.connect(connection_string)**: This line establishes a connection to the SQL Server using the connection string we defined earlier.
  - **with statement**: Ensures that the connection is automatically closed when the block of code completes, even if an error occurs, which prevents leaving unnecessary connections open.
- **Opening and Reading the SQL File**:
  - **open(file_path, 'r')**: This line opens the SQL file specified by file_path in read mode ('r').
  - **file.read()**: Reads the contents of the file (which should contain an SQL script) and stores it in the variable sql_script.
- **Executing the SQL Script**:
  - **conn.cursor()**: Creates a cursor object that allows us to execute SQL commands in the connected database.
  - **cursor.execute(sql_script)**: Executes the SQL script contained in the sql_script variable.
  - **with statement**: Just like the connection, the cursor is automatically closed after the SQL execution completes.
- **Success Message**:
  - If the SQL script runs successfully, this message is printed, indicating which script was executed. The os.path.basename(file_path) function extracts the file name from the full path to make the message more readable.

The function of error handling includes several except clauses to handle specific types of errors:

```
    except pyodbc.Error as e:
        print(f"Error executing {os.path.basename(file_path)}: {e}")
    except FileNotFoundError:
        print(f"File not found: {file_path}")
    except Exception as e:
        print(f"An error occurred with {os.path.basename(file_path)}: {e}")
```

Explanation

- **pyodbc.Error**:
  - This block catches any errors raised by pyodbc, which typically relate to issues with the database connection or SQL execution. It prints an error message along with the actual error message ({e}) returned by pyodbc.
- **FileNotFoundError**:
  - This block catches the case where the file specified by file_path doesn't exist. This helps users identify when the wrong file path is provided.
- **General Exception Handling**:

- This block catches any other exceptions not covered by the previous error types. It ensures that unexpected errors are still caught, and an appropriate error message is displayed.

Summary

In this step, we created a function to execute SQL files. The function handles errors, opens the file, connects to SQL Server, executes the script, and prints a success or error message. This allows for a robust and automated way of executing SQL files.

Step 6: Executing All SQL Files

Now that the function for executing SQL files is in place, we'll use it to loop through each SQL file in the directory and execute them one by one.

```
for sql_file in sql_files:
    sql_file_path = os.path.join(sql_directory, sql_file)
    execute_sql_file(sql_file_path)
```

Explanation

- **Loop through Each SQL File**:
    - This line initiates a loop that processes every item in the sql_files list.
    - **sql_files**: This is a list containing the names of SQL files, which we obtained in an earlier step.
    - **sql_file**: This variable holds the name of the current SQL file being processed in each iteration of the loop.

- **Create the Full Path for Each SQL File**:
    - **Purpose**: Creates the full file path for the current SQL file by combining the directory and the file name.
    - **os.path.join()**: Safely combines the directory and file names into a valid path that works across different operating systems.
    - **sql_directory**: Holds the path to the directory where the SQL files are stored.
    - **sql_file**: The name of the current SQL file.
    - **sql_file_path**: The result is the full path to the current SQL file, which is used to locate the file in the file system.

- **Execute the SQL File**:
    - **Purpose**: Calls the execute_sql_file() function, passing the full path of the current SQL file as an argument.

- **execute_sql_file()**: The function that opens, reads, and executes the SQL file on the database. The sql_file_path is passed to the function so it knows which file to process.

Summary

In this step, we looped through the list of SQL files, generating the full file path for each one and calling the execute_sql_file() function to execute the SQL script. This approach ensures that all SQL files in the directory are processed in sequence.

Final Output

After executing the Python code in this chapter, nine tables will be automatically created in SQL Server Management Studio (SSMS) under the "Automate Everything with Python" database, as shown in the examples provided. These tables will serve as structured storage for the data we'll be working with throughout the book. In the next chapter, we'll populate each of these tables with data, setting the foundation for more complex operations and analysis.

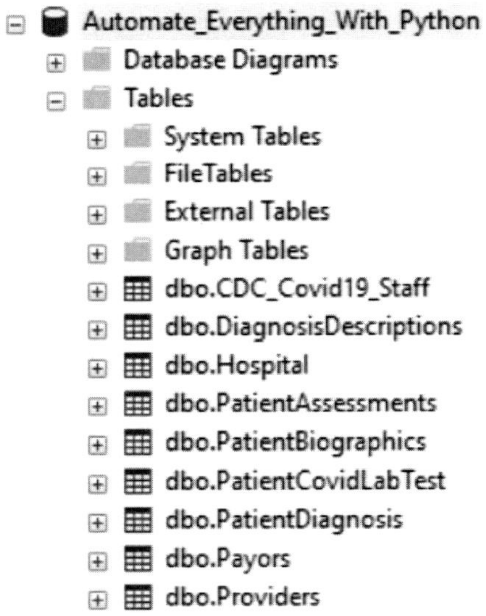

Integrating SQL execution within Python scripts provides a seamless way to automate database operations, allowing you to manage and manipulate data with precision and efficiency. By using Python to execute SQL files, you can:

- **Automate Data Handling**: Automatically create, update, and manage tables and records without manual intervention.
- **Ensure Consistency**: Repeat tasks consistently across various data sources and systems by running scripts that maintain standardized data structures.
- **Streamline Workflows**: Combine data processing and database management into a single workflow, reducing the need to switch between multiple tools.

- **Increase Flexibility**: Use Python's logic and data manipulation capabilities in tandem with SQL's powerful querying functions, enabling dynamic and flexible data operations.

As you progress, these automated SQL operations will become integral to building a scalable, efficient data automation pipeline.

Conclusion

Congratulations! You've now gained essential skills for automating SQL file execution with Python. Throughout this chapter, you learned how to efficiently connect to a SQL Server database, automate the process of reading and executing SQL files, and handle potential errors that may arise. By implementing this automation, you are streamlining workflows that typically require manual intervention, making your processes more efficient and less error-prone.

Here's a quick recap of the key skills you mastered in this chapter:

- **Connecting to SQL Server**: You established a connection to SQL Server using Python's pyodbc library, setting the foundation for automating SQL executions.
- **Listing and Filtering SQL Files**: You used Python to list and filter SQL files based on specific naming conventions, ensuring only relevant files were processed.
- **Automating SQL Execution**: You built a Python function to read SQL files, execute their contents, and handle errors with a robust try-except structure.
- **Error Handling and Feedback**: The function was designed to catch various types of errors (e.g., database connection issues, file not found) and provide informative feedback, ensuring smooth execution and debugging.

With this powerful tool, you can now automate repetitive SQL execution tasks, freeing up time and minimizing manual intervention. Keep refining your skills and continue exploring more ways to enhance your automation capabilities!

Chapter 3:
Loading Diverse Flat File Formats into SQL

Introduction: Mastering Flat File Imports into SQL

In this chapter, you will learn how to efficiently load data from a variety of flat file formats into a SQL Server database using Python. Flat files, such as CSVs, TXT files, and Excel spreadsheets, are common in data management, and understanding how to handle them is crucial for anyone working with databases. Throughout this chapter, we will explore five key sections that will equip you with the knowledge to handle diverse file types and formats, each posing its own challenges in terms of structure, data cleaning, and SQL integration.

Overview of the 5 Sections

1. **Loading CSV Files into SQL**:
   In this section, you will master importing CSV (Comma Separated Values) files into SQL Server. CSVs are widely used for data storage and sharing, making it critical to understand how to properly load and integrate them into databases. You'll learn how to read CSV files, manage data types like strings and numbers, and load this structured data into SQL.
2. **Loading TXT Files into SQL**:
   Here, we focus on loading TXT files, specifically tab-delimited ones, into SQL. TXT files can be a bit trickier due to non-standard delimiters and inconsistent data formats. You'll learn how to manipulate and load them into SQL while preserving key details like IDs and date formats.
3. **Loading Dirty TXT Files into SQL**:
   In real-world data management, you'll frequently encounter messy or "dirty" data. This section covers how to handle TXT files with extra rows, irrelevant columns, and non-standard formatting. You'll discover how to clean such files and load only the necessary data into SQL.
4. **Loading Excel Files into SQL**:
   Excel is one of the most common formats used in business environments. This section will guide you through extracting data from Excel spreadsheets, handling potential formatting issues, and loading data into SQL Server. You'll also learn how to manage specific data types like IDs, dates, and ZIP codes.
5. **Loading Excel Files from Specific Worksheets into SQL**:
   In the final section, we delve into selectively loading data from specific worksheets within an Excel file. You'll learn how to navigate multi-sheet workbooks, extract data from targeted sheets, and map it to SQL tables. This section is essential for handling complex Excel files with multiple datasets stored across different tabs.

How This Chapter Will Help You

By the end of this chapter, you will be equipped to:

- Handle various flat file formats (CSV, TXT, Excel) and navigate their unique challenges.

- Clean and transform "dirty" data before loading it into SQL.
- Manage and preserve important data types, such as dates, IDs, and numeric codes.
- Automate the process of importing large datasets into SQL, saving time and reducing manual errors.

This chapter is designed to give you practical skills that will be valuable in any database management role. Each section builds upon the previous one, progressively increasing your proficiency in handling increasingly complex file types. Let's dive into each section and master the art of importing flat files into SQL!

Chapter 3: Loading Diverse Flat File Formats into SQL

Section 1: Loading CSV Files into SQL

In this section, we will cover how to import data from CSV files into a SQL Server database using Python. CSV files are a common format for storing data, and mastering their import into SQL is an essential skill in data analysis and database management. Let’s walk through this process step-by-step.

Code Overview: Your Roadmap to Success

Before we start coding, let’s take a moment to understand the overall structure. The following screenshot illustrates what your script will look like once you've completed this section:

```python
# ==========================================================================
# Chapter 3 - Loading Diverse Flat File Formats into SQL.py
#
# Section 1: Load CSV file into SQL
# ==========================================================================

import pandas as pd
import pyodbc
from sqlalchemy import create_engine
import os

# Get the username
pc_username = os.getlogin()

data_directory = r"C:\\Users\\" + pc_username +  "\\Desktop\Automate_Everything_With_Python\\Data Files"


df_patient_diag = pd.read_csv(data_directory + '\\' + 'Patient Diagnosis.csv',
                              converters={'PatientID':str})

# Connect to SQL Server
conn_str = ('Driver={ODBC Driver 17 for SQL Server};'
            'Server=Enter_Your_Server_Name;' # Enter your server name here
            'Database=Automate_Everything_With_Python;'
            'UID=automation_user;'
            'PWD=AutomateEverything2024!;'
            'Trusted_Connection=no')

# Create a connection and cursor
conn = pyodbc.connect(conn_str)
cursor = conn.cursor()

# Create SQLAlchemy engine
engine = create_engine(f"mssql+pyodbc:///?odbc_connect={conn_str}")

# Define the table name
table_name = 'PatientDiagnosis'  # Change this to your desired table name

# Load the DataFrame to the SQL Server table
df_patient_diag.to_sql(table_name, engine, if_exists='replace', index=False)
```

This visual guide will help you see how each step connects to form a complete script for CSV file loading into SQL. Don't worry if it looks overwhelming now—by the end of this section, you'll be able to understand and write the code on your own.

Step 1: Importing Necessary Libraries

First, we need to import the libraries that will help us read CSV files, manipulate data, and communicate with the SQL Server.

```python
import pandas as pd
import pyodbc
from sqlalchemy import create_engine
import os
```

Explanation

- **pandas**: A powerful library for data manipulation and reading the CSV files.
- **pyodbc**: A module that allows Python to interact with SQL Server through ODBC connections.
- **sqlalchemy**: Simplifies database connections and SQL operations from within Python.
- **os**: Provides functions to interact with the operating system, such as file path handling.

By importing these libraries, we ensure that Python can handle both the file management and database operations needed for our task.

Step 2: Setting the SQL Directory Path

We will dynamically define the file path to where our CSV file is located on the desktop. This ensures that the script works for any user, regardless of the system setup.

```
# Get the username
pc_username = os.getlogin()

data_directory = r"C:\\Users\\" + pc_username +  "\\Desktop\Automate_Everything_With_Python\\Data Files"

```

Explanation

- **os.getlogin()**: Retrieves the username of the person logged into the system. This ensures that the path is personalized for each user.
- **data_directory**: Defines the full path where your CSV file is located by combining the user directory with the folder path.

This step ensures that the script can easily locate the file regardless of the user's environment.

Step 3: Reading the CSV File

Now, let's read the CSV file and load its contents into a pandas DataFrame, which is a flexible data structure ideal for data analysis.

```

df_patient_diag = pd.read_csv(data_directory + '\\' + 'Patient Diagnosis.csv',
                              converters={'PatientID':str})

```

Explanation

- **pd.read_csv()**: Reads the CSV file and converts it into a DataFrame.

- **converters={'PatientID': str}**: Ensures that the 'PatientID' column is treated as a string rather than a number. This is crucial when working with IDs that may contain leading zeros.

By specifying this converter, we prevent leading zeros from being stripped, preserving the accuracy of the patient IDs.

Step 4: Establishing the SQL Connection

Next, we establish a connection to the SQL Server by defining the necessary parameters in a connection string. Here's how to do that:

```
# Connect to SQL Server
conn_str = ('Driver={ODBC Driver 17 for SQL Server};'
            'Server=Enter_Your_Server_Name;' # Enter your server name here
            'Database=Automate_Everything_With_Python;'
            'UID=automation_user;'
            'PWD=AutomateEverything2024!;'
            'Trusted_Connection=no')

# Create a connection and cursor
conn = pyodbc.connect(conn_str)
cursor = conn.cursor()
```

Explanation

- **Driver**: Specifies the ODBC driver required to communicate with SQL Server.
- **Server**: The name of your SQL Server instance.
- **Database**: The name of the database where the data will be loaded.
- **UID and PWD**: The user credentials required to authenticate and connect to the SQL Server.
- **Trusted_Connection=no**: Specifies that we are not using Windows authentication.

This step establishes the connection and prepares the SQL Server for receiving data.

Step 5: Creating the SQLAlchemy Engine

The next step is to set up the SQLAlchemy engine, which simplifies communication between Python and SQL Server.

```
# Create SQLAlchemy engine
engine = create_engine(f"mssql+pyodbc:///?odbc_connect={conn_str}")

```

Explanation

- **create_engine()**: Creates a connection to the database using SQLAlchemy.
- **mssql+pyodbc**: Specifies the protocol for connecting to SQL Server via pyodbc.
- **odbc_connect={conn_str}**: Passes the connection string to the engine.

With this engine in place, we can now execute SQL operations from Python with ease.

Step 6: Loading Data into SQL

Finally, we can insert the data from the pandas DataFrame into the SQL Server database. Here's how:

```
# Define the table name
table_name = 'PatientDiagnosis'  # Change this to your desired table name

# Load the DataFrame to the SQL Server table
df_patient_diag.to_sql(table_name, engine, if_exists='replace', index=False)
```

Explanation

- **table_name**: Specifies the name of the SQL table where the data will be stored.
- **to_sql()**: This function writes the DataFrame to the specified SQL table.
- **if_exists='replace'**: Replaces the existing table with the new data. You can also use 'append' to add data to an existing table without overwriting it.
- **index=False**: Ensures that the DataFrame's index (row numbers) are not written to the SQL table.

This step efficiently loads the CSV data into the SQL database, making it queryable and structured.

Final Output

Once you've executed the code, the data from your CSV files will be successfully loaded into structured tables within SQL Server Management Studio (SSMS). Specifically, the data will be available in the **PatientDiagnosis** table, ready for analysis and reporting.

To view the data you just imported, follow these steps:

1. Open SSMS and navigate to the **PatientDiagnosis** table.
2. Right-click on the **PatientDiagnosis** table.
3. Select **"Select Top 1000 Rows"** from the context menu.

This will display the first 1,000 rows of data, allowing you to verify the import and inspect the contents of the table. You can quickly review your data to confirm that it has been accurately loaded.

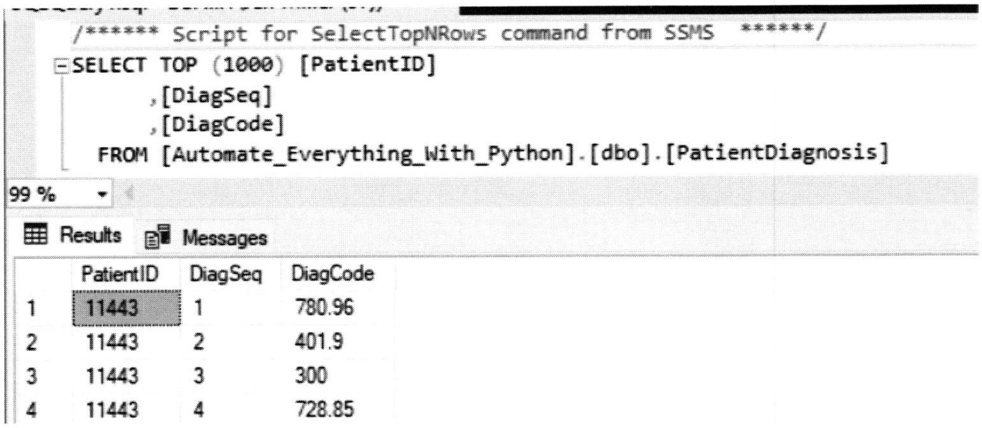

Powerful Data Extraction for Reporting

With this code, you've set up a streamlined workflow that enables data extraction from various files, loading them directly into SSMS. The structured data in **PatientDiagnosis** is now accessible for generating insightful reports, performing advanced queries, and supporting analytics—all within SSMS, a powerful tool for managing and analyzing data.

This automation lays the groundwork for efficient and error-free reporting, saving time and making SSMS a central resource for data-driven decision-making.

Conclusion

In this section, you've learned how to:

- Import necessary libraries for file handling and database interaction.
- Dynamically set file paths.
- Read CSV files using pandas.
- Establish a connection to SQL Server.
- Load data into SQL using SQLAlchemy.

With these skills, you can now automate the process of importing CSV data into SQL databases, ensuring data integrity and ease of access. These steps form a strong foundation for working with flat files, and as you continue to explore, you'll be able to handle even more complex file formats and scenarios.

Keep practicing, and soon you'll be able to seamlessly integrate CSV data into SQL databases as part of your everyday workflows!

Chapter 3: Loading Diverse Flat File Formats into SQL

Section 2: Loading TXT Files into SQL

In this section, we will cover how to import tab-delimited TXT files into a SQL Server database using Python. TXT files are another common format for storing data, and understanding how to efficiently read, process, and load them into SQL is essential for database management. Let's walk through this process step-by-step.

Code Overview: Your Blueprint for TXT-to-SQL Transformation

Before we start coding, let's take a moment to understand the overall structure. The following screenshot illustrates what your script will look like once you've completed this section:

```python
# Section 2: Load TXT file into SQL
#
# ============================================================================
import pandas as pd
import pyodbc
import os

# Get the username
pc_username = os.getlogin()

data_directory = r"C:\\Users\\" + pc_username +  "\\Desktop\Automate_Everything_With_Python\\Data Files"

df_patient_bio = pd.read_csv(data_directory + '\\' + 'Patient Biographics.txt',
                             sep='\t',
                             converters={'PatientID':str,
                                         'InsuranceID':str})

# Convert date object to datetime
df_patient_bio['BirthDate'] = pd.to_datetime(df_patient_bio['BirthDate'],format='%m/%d/%Y')

# Server Connection
conn_str = ('Driver={ODBC Driver 17 for SQL Server};'
            'Server=Enter_Your_Server_Name;' # Enter your server name here
            'Database=Automate_Everything_With_Python;'
            'UID=automation_user;'
            'PWD=AutomateEverything2024!;'
            'Trusted_Connection=no')

conn = pyodbc.connect(conn_str)
cursor = conn.cursor()

# Insert DataFrame/csv file into SQL table
for row in df_patient_bio.itertuples():
    cursor.execute('''insert into PatientBiographics (PatientID,PatientName,BirthDate,MaritalStatus,Race,Gender,Phone,InsuranceID)
                   values(?,?,?,?,?,?,?,?)''',
                   row.PatientID,
                   row.PatientName,
                   row.BirthDate,
                   row.MaritalStatus,
                   row.Race,
                   row.Gender,
                   row.Phone,
                   row.InsuranceID)

conn.commit()

# Close Connection
conn.close()
```

This visual guide will help you see how each step connects to form a complete script for loading TXT files into SQL. By the end of this section, you'll understand the code and be able to build your own script.

Step 1: Importing the Necessary Libraries

We begin by importing the libraries that will help us read TXT files, manipulate data, and communicate with SQL Server.

```python
import pandas as pd
import pyodbc
import os
```

Explanation

- **pandas**: A versatile library for data manipulation and reading the TXT files.
- **pyodbc**: Allows Python to interact with SQL Server through ODBC connections.
- **os**: Provides functions to interact with the operating system, such as file path handling.

These libraries will be critical in processing the TXT file and loading the data into SQL Server.

Step 2: Reading the TXT File

Next, we load the tab-delimited TXT file into a pandas DataFrame, ensuring that certain columns are treated as strings to preserve any leading zeros.

```
# Get the username
pc_username = os.getlogin()

data_directory = r"C:\\Users\\" + pc_username +  "\\Desktop\Automate_Everything_With_Python\\Data Files"

df_patient_bio = pd.read_csv(data_directory + '\\' + 'Patient Biographics.txt',
                             sep='\t',
                             converters={'PatientID':str,
                                         'InsuranceID':str})
```

Explanation

- **os.getlogin()**: Retrieves the username of the person logged into the system. This ensures that the file path works for any user.
- **pd.read_csv()**: Reads the tab-delimited TXT file into a DataFrame.
- **sep='\t'**: Specifies that the file is tab-delimited.
- **converters={'PatientID': str, 'InsuranceID': str}**: Ensures that the 'PatientID' and 'InsuranceID' columns are treated as strings, preserving any leading zeros or special characters.

This step prepares the data for further processing by loading it into a DataFrame.

Step 3: Date Conversion

Handling dates correctly is critical in most real-world data. Let's convert the 'BirthDate' column to datetime objects to make it easier to work with in SQL.

```
# Convert date object to datetime
df_patient_bio['BirthDate'] = pd.to_datetime(df_patient_bio['BirthDate'],format='%m/%d/%Y')
```

Explanation

- **pd.to_datetime()**: Converts the 'BirthDate' column from strings to datetime objects.
- **format='%m/%d/%Y'**: Specifies the format of the date in the TXT file (Month/Day/Year).

Quick Date Format Reference:

- '%m/%d/%Y': Month/Day/Year (e.g. 12/25/2023)
- '%d-%m-%Y': Day-Month-Year (e.g. 25-12-2023)
- '%Y/%m/%d': Year/Month/Day (e.g. 2023/12/25)

Using the correct date format ensures that the data is accurately loaded into the SQL database.

Step 4: Establishing the SQL Connection

Now, we will establish a connection to the SQL Server using the following connection string:

```
# Server Connection
conn_str = ('Driver={ODBC Driver 17 for SQL Server};'
            'Server=Enter_Your_Server_Name;' # Enter your server name here
            'Database=Automate_Everything_With_Python;'
            'UID=automation_user;'
            'PWD=AutomateEverything2024!;'
            'Trusted_Connection=no')

conn = pyodbc.connect(conn_str)
cursor = conn.cursor()
```

Explanation

- **Driver**: Specifies the ODBC driver required to communicate with SQL Server.
- **Server**: The name of the SQL Server instance.
- **Database**: The name of the SQL database.
- **UID and PWD**: The user credentials for authentication.
- **Trusted_Connection=no**: Specifies that Windows authentication is not used.

This connection prepares SQL Server to receive the data.

Step 5: Inserting Data Row by Row

Now that the data is ready, we can insert it into the SQL table row by row using a loop.

```
# Insert DataFrame/csv file into SQL table
for row in df_patient_bio.itertuples():
    cursor.execute('''insert into PatientBiographics (PatientID,PatientName,BirthDate,MaritalStatus,Race,Gender,Phone,InsuranceID)
                values(?,?,?,?,?,?,?,?)''',
                row.PatientID,
                row.PatientName,
                row.BirthDate,
                row.MaritalStatus,
                row.Race,
                row.Gender,
                row.Phone,
                row.InsuranceID)
```

Explanation

- **itertuples()**: Loops through each row of the DataFrame efficiently.
- **cursor.execute()**: Executes the SQL query to insert data into the PatientBiographics table. The ? placeholders are filled with corresponding data from each row.

This step sends the data to the SQL Server row by row.

Step 6: Committing Changes and Closing the Connection

After inserting all the data, it's important to commit the changes and close the connection to the database.

```
conn.commit()

# Close Connection
conn.close()
```

Explanation

- **conn.commit()**: Saves all the changes made during the session to the SQL database.
- **conn.close()**: Closes the connection to SQL Server, releasing any resources used.

Final Output

Once you've executed the code, the data from your CSV files will be successfully loaded into structured tables within SQL Server Management Studio (SSMS). Specifically, the data will be available in the **PatientBiographics** table, ready for analysis and reporting.

To view the data you just imported, follow these steps:

1. Open SSMS and navigate to the **PatientBiographics** table.

2. Right-click on the **PatientBiographics** table.
3. Select **"Select Top 1000 Rows"** from the context menu.

This will display the first 1,000 rows of data, allowing you to verify the import and inspect the contents of the table. You can quickly review your data to confirm that it has been accurately loaded.

```
SELECT TOP (1000) [PatientID]
      ,[PatientName]
      ,[BirthDate]
      ,[MaritalStatus]
      ,[Race]
      ,[Gender]
      ,[Phone]
      ,[InsuranceID]
  FROM [Automate_Everything_With_Python].[dbo].[PatientBiographics]
```

	PatientID	PatientName	BirthDate	MaritalStatus	Race	Gender	Phone	InsuranceID
1	11000	Jon, Yang	1971-10-06	M	P	M	1 (11) 500 555-0162	350
2	11006	Janet, Alvarez	1976-12-02	S	H	F	1 (11) 500 555-0184	210
3	11014	Sydney, Bennett	1973-11-06	S	E	F	431-555-0156	350

Powerful Data Extraction for Reporting

With this code, you've set up a streamlined workflow that enables data extraction from various files, loading them directly into SSMS. The structured data in **PatientBiographics** is now accessible for generating insightful reports, performing advanced queries, and supporting analytics all within SSMS, a powerful tool for managing and analyzing data.

This automation lays the groundwork for efficient and error-free reporting, saving time and making SSMS a central resource for data-driven decision-making.

Conclusion

Congratulations! You've successfully learned how to load tab-delimited TXT files into SQL using Python. Here's what you've accomplished:

- Imported necessary libraries to handle TXT files and interact with SQL Server.
- Loaded a TXT file into a pandas DataFrame and ensured proper handling of strings and dates.
- Established a connection to SQL Server.
- Inserted data row by row into a SQL table.
- Committed changes and closed the connection.

As you continue to work with TXT files, keep these tips in mind:

- Always validate your data, especially when working with dates or sensitive fields.
- Adjust date formats as needed to match the structure of your data.

With these skills, you're now well-prepared to handle TXT files and transform them into structured SQL tables. Keep practicing, and soon you'll be proficient in integrating flat files into SQL databases with ease!

Chapter 3: Loading Diverse Flat File Formats into SQL

Section 3: Loading "Dirty" TXT Files into SQL

In this section, we'll tackle a common challenge in data management: handling messy or "dirty" data from TXT files and efficiently loading it into a SQL Server database. Real-world data often contains inconsistencies, extra information, or non-standard formatting, and by the end of this section, you'll have learned how to clean and load such data into SQL.

Code Overview: Preparing for Data Cleansing

Before diving into the code, let's take a quick look at the overall process. Here's a visual representation of the complete script you'll be building:

```
# ============================================================================
#
# Section 3: Load Dirty TXT file into SQL
#
# ============================================================================
import pandas as pd
import pyodbc
import os

# Get the username
pc_username = os.getlogin()

data_directory = r"C:\\Users\\" + pc_username +  "\\Desktop\Automate_Everything_With_Python\\Data Files"

df_payor = pd.read_csv(data_directory + '\\' + 'Payor.txt',
                       sep='\t',
                       skiprows=3,
                       usecols=[1,2],
                       converters={1:str})

# Set the first row as the header
df_payor.columns = df_payor.iloc[0]

# Drop the first row as it's now the header
df_payor = df_payor.drop(df_payor.index[0])

conn_str = ('Driver={ODBC Driver 17 for SQL Server};'
            'Server=Enter_Your_Server_Name;' # Enter your server name here
            'Database=Automate_Everything_With_Python;'
            'UID=automation_user;'
            'PWD=AutomateEverything2024!;'
            'Trusted_Connection=no')

conn = pyodbc.connect(conn_str)
cursor = conn.cursor()

# Insert DataFrame/csv file into SQL table
for row in df_payor.itertuples():
    cursor.execute('''insert into Payors (PayorID,Payor)
                   values(?,?)''',
                   row.PayorID,
                   row.Payor)

conn.commit()

# Close Connection
conn.close()
```

By breaking the task into smaller steps, this structure will help guide you through the cleaning and loading of the messy data into SQL.

Step 1: Importing Necessary Libraries

We will need several libraries to handle file processing, data cleaning, and SQL interaction.

```python
import pandas as pd
import pyodbc
import os
```

Explanation

- **pandas**: A powerful data manipulation library for reading and cleaning TXT files.
- **pyodbc**: Facilitates communication between Python and SQL Server, allowing us to send SQL queries.
- **os**: Provides functions to dynamically access files and directories based on the system environment.

These libraries equip us with the necessary tools to read the TXT file, clean the data, and insert it into SQL.

Step 2: Reading the "Dirty" TXT File

Now we need to load the "dirty" TXT file, which may contain extra rows, unnecessary columns, and non-standard formatting. To handle this, we will use specific parameters in pandas to clean the data during the loading process.

```python
# Get the username
pc_username = os.getlogin()

data_directory = r"C:\\Users\\" + pc_username +  "\\Desktop\Automate_Everything_With_Python\\Data Files"

df_payor = pd.read_csv(data_directory + '\\' + 'Payor.txt',
                       sep='\t',
                       skiprows=3,
                       usecols=[1,2],
                       converters={1:str})

# Set the first row as the header
df_payor.columns = df_payor.iloc[0]

# Drop the first row as it's now the header
df_payor = df_payor.drop(df_payor.index[0])
```

Explanation

- **os.getlogin()**: Retrieves the username of the person logged into the system. This ensures that the file path works for any user.
- **pd.read_csv()**: Reads the TXT file into a pandas DataFrame, similar to how we handle CSV files but with extra cleaning features.
- **sep='\t'**: Specifies that the file is tab-delimited, meaning data fields are separated by tabs instead of commas.
- **skiprows=3**: Skips the first three rows, which contains irrelevant information or metadata. This is useful for ignoring non-data content at the top of the file.

- **usecols=[1,2]**: Selects only the necessary column (in this case, columns 1 and 2) from the file. This helps avoid loading unnecessary data into memory.
- **converters={1: str}**: Ensures that the second column (index 1) is treated as a string, preserving any leading zeros or special characters that could be important for identifiers like IDs or zip codes.
- `df_payor.columns = df_payor.iloc[0]`: Sets the first row of the DataFrame as the header. This is often needed when the first row contains column names but is mistakenly read as data.
- `df_payor = df_payor.drop(df_payor.index[0])`: Drops the first row after it has been set as the header, so it doesn't appear twice in the DataFrame. This cleans up the dataset by removing redundant information.

This step cleans up the "dirty" data by skipping irrelevant rows, selecting essential columns, and ensuring data types are preserved, and correctly setting the header.

Step 3: Establishing the SQL Connection

Next, we establish a connection to SQL Server, which allows Python to interact with the database. This connection is necessary for loading the cleaned data into a SQL table.

```
conn_str = ('Driver={ODBC Driver 17 for SQL Server};'
            'Server=Enter_Your_Server_Name;' # Enter your server name here
            'Database=Automate_Everything_With_Python;'
            'UID=automation_user;'
            'PWD=AutomateEverything2024!;'
            'Trusted_Connection=no')

conn = pyodbc.connect(conn_str)
cursor = conn.cursor()
```

Explanation

- **Driver**: Specifies the ODBC driver for SQL Server communication.
- **Server**: Name of the SQL Server instance to connect to.
- **Database**: Specifies the database where the cleaned data will be loaded.
- **UID and PWD**: The user credentials for SQL Server authentication. These allow Python to log in to the SQL Server.
- **Trusted_Connection=no**: Specifies that Windows authentication is not used, and SQL Server credentials (UID/PWD) are provided instead.

This step prepares SQL Server for receiving the cleaned data by creating a connection through the ODBC driver.

Step 4: Inserting Data into SQL

Now that the data is cleaned and we've established a connection to SQL Server, we can insert the data into the database. This process is similar to previous sections but uses the cleaned DataFrame from the TXT file.

```
# Insert DataFrame/csv file into SQL table
for row in df_payor.itertuples():
    cursor.execute('''insert into Payors (PayorID,Payor)
                    values(?,?)''',
                    row.PayorID,
                    row.Payor)

conn.commit()
```

Explanation

- **itertuples()**: Efficiently iterates over each row of the DataFrame. Each row is represented as a tuple, allowing easy access to its data.
- **cursor.execute()**: Executes a SQL query to insert the data from the DataFrame into the Payor table. The ? placeholders are replaced with the values from the PayorID and PayorName fields of each row.

This loop processes each row of the cleaned data and inserts it into the SQL table, row by row, ensuring that the data is properly loaded into the database.

Step 5: Committing Changes and Closing the Connection

Once all the data has been inserted into SQL, it is important to commit the changes to make sure they are saved and then close the connection to release resources.

```
conn.commit()

# Close Connection
conn.close()

```

Explanation

- **conn.commit()**: Saves all the changes made to the SQL database during the session, ensuring that the data insertions are permanent.
- **conn.close()**: Closes the connection to SQL Server, freeing up resources and ensuring that no further changes are made to the database.

This final step ensures that the data is safely saved to the SQL database and the connection is properly closed.

Final Output

Once you've executed the code, the data from your CSV files will be successfully loaded into structured tables within SQL Server Management Studio (SSMS). Specifically, the data will be available in the **Payors** table, ready for analysis and reporting.

To view the data you just imported, follow these steps:

1. Open SSMS and navigate to the **Payors** table.
2. Right-click on the **Payors** table.
3. Select **"Select Top 1000 Rows"** from the context menu.

This will display the first 1,000 rows of data, allowing you to verify the import and inspect the contents of the table. You can quickly review your data to confirm that it has been accurately loaded.

```
SELECT TOP (1000) [PayorID]
      ,[Payor]
  FROM [Automate_Everything_With_Python].[dbo].[Payors]
```

99 %

Results | Messages

	PayorID	Payor
1	1	SELF PAY
2	197	SELF PAY MEDICAID PENDING
3	200	BCBS PPO TEXAS

Powerful Data Extraction for Reporting

With this code, you've set up a streamlined workflow that enables data extraction from various files, loading them directly into SSMS. The structured data in **Payors** is now accessible for generating insightful reports, performing advanced queries, and supporting analytics—all within SSMS, a powerful tool for managing and analyzing data.

This automation lays the groundwork for efficient and error-free reporting, saving time and making SSMS a central resource for data-driven decision-making.

Conclusion

Congratulations! You've successfully tackled the challenge of loading "dirty" TXT files into SQL using Python. Here's a summary of what you've learned:

- **Data Cleaning**: You handled non-standard data formats by skipping irrelevant rows and selecting necessary columns during the data loading process.
- **SQL Connection**: You established a connection to SQL Server using Python and inserted cleaned data into a SQL table efficiently.
- **Error Prevention**: By using pandas advanced parameters like skiprows, usecols, and converters, you tailored your data imports to your specific needs.

Key Takeaways:

- Always inspect the structure of your data before loading it into SQL. Techniques like skipping unnecessary rows and selecting relevant columns streamline your workflow.
- Previewing large files before loading them into SQL helps identify potential issues early on, reducing errors during processing.
- Use Python's pandas and pyodbc libraries together to manage and insert data into SQL databases effectively.

With these skills, you're well-prepared to clean and transform even the messiest datasets, turning chaotic data into well-structured SQL tables. Keep practicing, and soon you'll master the art of handling dirty data in SQL workflows!

Chapter 3: Loading Diverse Flat File Formats into SQL

Section 4: Loading Excel Files into SQL

In this section, we will explore how to efficiently load data from Excel files into a SQL Server database using Python. Excel is one of the most common formats used in business environments, and handling data from complex Excel sheets is a vital skill. By the end of this section, you'll be able to extract and load data from Excel spreadsheets, whether they have single or multiple sheets, into structured SQL databases.

Code Overview: Your Guide for Excel-to-SQL Transformation

Before diving into the code, let's take a moment to review what the final script will look like:

```
# ============================================================================
#
# Section 4: Load Excel file into SQL
#
# ============================================================================
import pandas as pd
import pyodbc
import os

# Get the username
pc_username = os.getlogin()

data_directory = r"C:\\Users\\" + pc_username +  "\\Desktop\Automate_Everything_With_Python\\Data Files"

df_hospital = pd.read_excel(data_directory + '\\' + 'Hospitals.xlsx',
                            skiprows=2,
                            converters={'ClinicID':str,
                                        'Zip':str})[['ClinicID','ClinicName','City','State','Zip']]

conn_str = ('Driver={ODBC Driver 17 for SQL Server};'
            'Server=Enter_Your_Server_Name;' # Enter your server name here
            'Database=Automate_Everything_With_Python;'
            'UID=automation_user;'
            'PWD=AutomateEverything2024!;'
            'Trusted_Connection=no')

conn = pyodbc.connect(conn_str)
cursor = conn.cursor()

# Insert DataFrame/csv file into SQL table
for row in df_hospital.itertuples():
    cursor.execute('''insert into Hospital (ClinicID,ClinicName,City,State,Zip)
                    values(?,?,?,?,?)''',
                    row.ClinicID,
                    row.ClinicName,
                    row.City,
                    row.State,
                    row.Zip)

conn.commit()

# Close Connection
conn.close()
```

This roadmap will help you see how each step connects, from reading the Excel file to inserting the data into SQL. Don't worry if it looks overwhelming now. By the end of this section, you'll be able to understand and write this code yourself.

Step 1: Importing Necessary Libraries

Let's first ensure that the essential libraries are imported. If you've followed along with the previous sections, these libraries should already be in place.

```python
import pandas as pd
import pyodbc
import os
```

Explanation

- **pandas**: A library for data manipulation, which allows us to read and handle Excel files easily.
- **pyodbc**: Enables Python to communicate with SQL Server.
- **os**: Provides functions to interact with the operating system, such as handling file paths dynamically.

These libraries are crucial for reading Excel data and interacting with SQL databases.

Step 2: Reading the Excel File

We'll now read the Excel file using pandas, which makes this process simple and effective.

```python
# Get the username
pc_username = os.getlogin()

data_directory = r"C:\\Users\\" + pc_username +  "\\Desktop\Automate_Everything_With_Python\\Data Files"

df_hospital = pd.read_excel(data_directory + '\\' + 'Hospitals.xlsx',
                            skiprows=2,
                            converters={'ClinicID':str,
                                        'Zip':str})[['ClinicID','ClinicName','City','State','Zip']]

```

Explanation

- **pd.read_excel()**: Reads the Excel file and loads it into a pandas DataFrame.
- **skiprows=2**: Skips the first two rows, which may contain irrelevant information or titles that you don't need.
- **converters={'ClinicID': str, 'Zip': str}**: Ensures that the 'ClinicID' and 'Zip' columns are treated as strings, which is crucial for maintaining leading zeros.
- **[['ClinicID', 'ClinicName', 'City', 'State', 'Zip']]**: Selects only the relevant columns from the Excel file, ensuring that the DataFrame contains only the data we need.

Step 3: Inserting Data into SQL

Once the data is loaded into the DataFrame, we're ready to insert it into the SQL database. Here's the code to do that:

```python
conn_str = ('Driver={ODBC Driver 17 for SQL Server};'
            'Server=Enter_Your_Server_Name;' # Enter your server name here
            'Database=Automate_Everything_With_Python;'
            'UID=automation_user;'
            'PWD=AutomateEverything2024!;'
            'Trusted_Connection=no')

conn = pyodbc.connect(conn_str)
cursor = conn.cursor()

# Insert DataFrame/csv file into SQL table
for row in df_hospital.itertuples():
    cursor.execute('''insert into Hospital (ClinicID,ClinicName,City,State,Zip)
                    values(?,?,?,?,?)''',
                    row.ClinicID,
                    row.ClinicName,
                    row.City,
                    row.State,
                    row.Zip)

conn.commit()

# Close Connection
conn.close()
```

Explanation

- **conn_str**: This connection string sets up communication between Python and SQL Server using credentials you created earlier.
- **cursor.execute()**: Executes SQL queries to insert data row by row into the Hospital table. The ? placeholders are filled with data from the corresponding columns in each row.
- **conn.commit()**: Commits the transaction, saving the changes to the SQL database.
- **conn.close()**: Closes the connection to SQL Server.

This code sends the hospital data from the DataFrame to the SQL table, ensuring all rows are successfully inserted.

Final Output

Once you've executed the code, the data from your CSV files will be successfully loaded into structured tables within SQL Server Management Studio (SSMS). Specifically, the data will be available in the **Hospital** table, ready for analysis and reporting.

To view the data you just imported, follow these steps:

1. Open SSMS and navigate to the **Hospital** table.

2. Right-click on the **Hospital** table.
3. Select **"Select Top 1000 Rows"** from the context menu.

This will display the first 1,000 rows of data, allowing you to verify the import and inspect the contents of the table. You can quickly review your data to confirm that it has been accurately loaded.

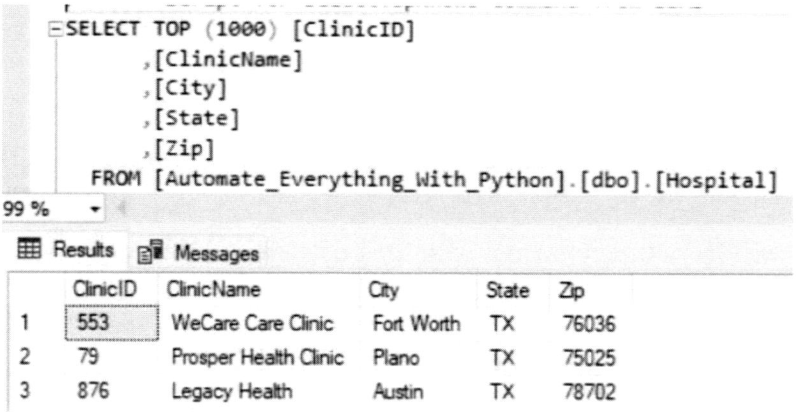

Powerful Data Extraction for Reporting

With this code, you've set up a streamlined workflow that enables data extraction from various files, loading them directly into SSMS. The structured data in **Hospital** is now accessible for generating insightful reports, performing advanced queries, and supporting analytics—all within SSMS, a powerful tool for managing and analyzing data.

This automation lays the groundwork for efficient and error-free reporting, saving time and making SSMS a central resource for data-driven decision-making.

Conclusion

Congratulations! You've successfully learned how to transform Excel files into structured SQL databases using Python. In this section, you covered:

- Reading data from single or multiple Excel sheets.
- Handling specific columns and data types, especially for IDs and ZIP codes.
- Inserting data into SQL databases with ease.

Tips to Keep in Mind:

- Always inspect your Excel data before and after processing to ensure accuracy.
- Be mindful of data types, especially when working with IDs, ZIP codes, or dates.

- For large Excel files, consider reading the data in chunks to manage memory more efficiently.

With these skills, you can efficiently transform even the most complex Excel files into organized SQL databases. Keep practicing, and soon you'll master the process of Excel-to-SQL transformation!

Chapter 3: Loading Diverse Flat File Formats into SQL

Section 5: Loading Excel Files into SQL from Specific Worksheets

In this section, we will learn how to selectively load data from specific worksheets within an Excel workbook into a SQL Server database. Excel workbooks often contain multiple sheets, each storing different types of data. By the end of this section, you'll be equipped to navigate multi-sheet Excel files and efficiently load only the relevant data into SQL.

Code Overview: Your Blueprint for Selective Excel-to-SQL Transformation

Before we dive into the code, here's an overview of the final script you'll be building:

```
# ==========================================================================
#
# Section 5: Load Excel files into SQL from specific worksheet
#
# ==========================================================================
import pandas as pd
import pyodbc
import os

# Get the username
pc_username = os.getlogin()

data_directory = r"C:\\Users\\" + pc_username +  "\\Desktop\Automate_Everything_With_Python\\Data Files"

conn_str = ('Driver={ODBC Driver 17 for SQL Server};'
            'Server=Enter_Your_Server_Name;' # Enter your server name here
            'Database=Automate_Everything_With_Python;'
            'UID=automation_user;'
            'PWD=AutomateEverything2024!;'
            'Trusted_Connection=no')

conn = pyodbc.connect(conn_str)
cursor = conn.cursor()

# Load Providers worksheet from excel file
df_providers = pd.read_excel(data_directory + '\\' + 'Raw Data.xlsx',
                          sheet_name='Providers',
                          converters={'ProviderID':str})

# Load Patient Assessments worksheet from excel file
df_patientassessments = pd.read_excel(data_directory + '\\' + 'Raw Data.xlsx',
                          sheet_name='Patient Assessments',
                          converters={'ProviderID':str,
                                      'ClinicID':str,
                                      'PatientID':str})

```

```
# Insert DataFrame into SQL table
for row in df_providers.itertuples():
    cursor.execute('''insert into Providers (ProviderID,ProviderName,Phone)
                    values(?,?,?)''',
                    row.ProviderID,
                    row.ProviderName,
                    row.Phone)

# Insert DataFrame into SQL table
for row in df_patientassessments.itertuples():
    cursor.execute('''insert into PatientAssessments (PostPeriod,DateEntered,ProviderID,ClinicID,
                   DateTimeAssessment,PatientID,PatientName,Temperature,Pulse_Oximeter,SOB,Cough,
                   Abdominal_Pain,Diarrhea_Or_Other_Gi_Upset,Nausea,Loss_Of_Taste,
                   Red_Shadowed_Eyes_Or_Pink_Eye,Tingling_Sensation_Of_Face_Or_Hands,Sore_Throat,
                   Chills_And_Or_Repeated_Shaking_With_Chills,Muscle_Pain,Loss_Of_Smell,Headache)
                    values(?,?,?,?,?,?,?,?,?,?,?,?,?,?,?,?,?,?,?,?,?,?)''',
                    row.PostPeriod,row.DateEntered,row.ProviderID,row.ClinicID,row.DateTimeAssessment,
                    row.PatientID,row.PatientName,row.Temperature,row.Pulse_Oximeter,row.SOB,row.Cough,
                    row.Abdominal_Pain,row.Diarrhea_Or_Other_Gi_Upset,row.Nausea,row.Loss_Of_Taste,
                    row.Red_Shadowed_Eyes_Or_Pink_Eye,row.Tingling_Sensation_Of_Face_Or_Hands,
                    row.Sore_Throat,row.Chills_And_Or_Repeated_Shaking_With_Chills,
                    row.Muscle_Pain,row.Loss_Of_Smell,row.Headache)

conn.commit()

# Close Connection
conn.close()
```

This script covers everything from identifying specific worksheets to mapping worksheet columns to SQL fields. It ensures that only the necessary data is loaded into the SQL database, allowing for cleaner and more efficient data management.

Step 1: Importing the Necessary Libraries and Setting Up the Environment

To start, we need to import the libraries that will allow us to read Excel files and interact with SQL Server.

```
import pandas as pd
import pyodbc
import os
```

Explanation

- **pandas**: For reading Excel files and manipulating data.
- **pyodbc**: For establishing a connection to SQL Server and executing SQL queries.
- **os**: To dynamically access file paths based on the current user's profile.

This step sets up the basic environment and ensures that the file path is dynamically generated to fit the user's system.

Step 2: Establishing the SQL Connection

Next, we set up the connection to the SQL Server.

```
# Get the username
pc_username = os.getlogin()

data_directory = r"C:\\Users\\" + pc_username +  "\\Desktop\Automate_Everything_With_Python\\Data Files"

conn_str = ('Driver={ODBC Driver 17 for SQL Server};'
            'Server=ECHO1\SQLEXPRESS;' # Enter your server name here
            'Database=Automate_Everything_With_Python;'
            'UID=automation_user;'
            'PWD=AutomateEverything2024!;'
            'Trusted_Connection=no')

conn = pyodbc.connect(conn_str)
cursor = conn.cursor()
```

Explanation

- **conn_str**: Contains all the necessary parameters to connect to the SQL Server, including the server name, database name, and user credentials.
- **cursor**: Used to execute SQL commands.

This connection prepares SQL Server for inserting data from Excel.

Step 3: Reading Specific Excel Worksheets

In this section, we efficiently load and convert data from specific worksheets within the "*Raw Data.xlsx*" workbook located in the "Data Files" folder. By using the pd.read_excel() function with targeted parameters, we can read worksheets like 'Providers' and 'Patient Assessments,' and ensure crucial IDs are consistently handled as strings to avoid data type conflicts. This approach provides a structured way to manage data, making analysis and automation smoother and more reliable.

```
# Load Providers worksheet from excel file
df_providers = pd.read_excel(data_directory + '\\' + 'Raw Data.xlsx',
                             sheet_name='Providers',
                             converters={'ProviderID':str})

# Load Patient Assessments worksheet from excel file
df_patientassessments = pd.read_excel(data_directory + '\\' + 'Raw Data.xlsx',
                             sheet_name='Patient Assessments',
                             converters={'ProviderID':str,
                                         'ClinicID':str,
                                         'PatientID':str})
```

Explanation

Provider Data Extraction:

- **pd.read_excel()**: Reads the Excel file and loads the data into a pandas DataFrame.
- **sheet_name='Providers'**: Specifies the worksheet from which to read data.
- **converters={'ProviderID': str}**: Ensures that the 'ProviderID' column is treated as a string, avoiding potential issues with data types.

Patient Assessments Data Extraction:

- **pd.read_excel()**: Reads the Excel file and loads the data into a pandas DataFrame.
- **sheet_name='Patient Assessments'**: Specifies the worksheet from which to read data.
- **converters={'ProviderID': str, 'ClinicID': str, 'PatientID': str}**: Ensures that the 'ProviderID,' 'ClinicID,' and 'PatientID' columns are treated as strings, avoiding potential issues with data types.

This step extracts data from the "Raw Data.xlsx" workbook and the specified worksheets and prepares it for further processing.

Step 4: Inserting Data into SQL

This section iterates through each row in the DataFrames and insert data into SQL tables. Using a loop with itertuples() for efficient row access, we can map each DataFrame column to corresponding SQL table columns, ensuring data is properly stored and accessible for further analysis and use.

```python
# Insert DataFrame into SQL table
for row in df_providers.itertuples():
    cursor.execute('''insert into Providers (ProviderID,ProviderName,Phone)
                    values(?,?,?)''',
                    row.ProviderID,
                    row.ProviderName,
                    row.Phone)

# Insert DataFrame into SQL table
for row in df_patientassessments.itertuples():
    cursor.execute('''insert into PatientAssessments (PostPeriod,DateEntered,ProviderID,ClinicID,
                   DateTimeAssessment,PatientID,PatientName,Temperature,Pulse_Oximeter,SOB,Cough,
                   Abdominal_Pain,Diarrhea_Or_Other_Gi_Upset,Nausea,Loss_Of_Taste,
                   Red_Shadowed_Eyes_Or_Pink_Eye,Tingling_Sensation_Of_Face_Or_Hands,Sore_Throat,
                   Chills_And_Or_Repeated_Shaking_With_Chills,Muscle_Pain,Loss_Of_Smell,Headache)
                    values(?,?,?,?,?,?,?,?,?,?,?,?,?,?,?,?,?,?,?,?,?,?)''',
                    row.PostPeriod,row.DateEntered,row.ProviderID,row.ClinicID,row.DateTimeAssessment,
                    row.PatientID,row.PatientName,row.Temperature,row.Pulse_Oximeter,row.SOB,row.Cough,
                    row.Abdominal_Pain,row.Diarrhea_Or_Other_Gi_Upset,row.Nausea,row.Loss_Of_Taste,
                    row.Red_Shadowed_Eyes_Or_Pink_Eye,row.Tingling_Sensation_Of_Face_Or_Hands,
                    row.Sore_Throat,row.Chills_And_Or_Repeated_Shaking_With_Chills,
                    row.Muscle_Pain,row.Loss_Of_Smell,row.Headache)
```

Explanation

Provider Data Insertion:

- **for row in df_providers.itertuples()**: Iterates through each row in the df_providers DataFrame as tuples, enabling easy access to individual column values.
- **cursor.execute()**: Executes a SQL INSERT statement for each row.
- **SQL Statement**: Inserts data into the 'Providers' table, mapping columns to the values (ProviderID, ProviderName, Phone) from each row.
- **Values(?)**: The ? marks are placeholders for values that will be provided in the same order as they appear in the SQL statement. Each ? corresponds to a value from the row in the DataFrame.
- **row.ProviderID, row.ProviderName, row.Phone**: Maps each specific DataFrame column value to the corresponding SQL table column, ensuring data integrity during insertion.

Patient Assessments Data Insertion:

- **for row in df_patientassessments.itertuples()**: Iterates through each row in the df_patientassessments DataFrame as tuples.
- **cursor.execute()**: Executes an SQL INSERT statement for each row in the DataFrame.
- **SQL Statement**: Inserts data into the 'PatientAssessments' table, mapping each specific DataFrame column to a SQL table column.
- **Values(?)**: The ? marks are placeholders for values that will be provided in the same order as they appear in the SQL statement. Each ? corresponds to a value from the row in the DataFrame.
- **Multiple columns**: Maps each DataFrame column (e.g., row.PostPeriod, row.DateEntered, etc.) to corresponding SQL columns like PostPeriod, DateEntered,

ProviderID, etc., ensuring comprehensive data capture across all relevant patient assessment attributes.

Step 7: Committing Changes and Closing the Connection

Once all data has been inserted, it's essential to commit the changes and close the connection to the SQL Server.

```
conn.commit()

# Close Connection
conn.close()

```

Explanation

- **conn.commit()**: Saves the changes to the SQL database.
- **conn.close()**: Closes the connection, ensuring that resources are released properly.

Final Output

Once you've executed the code, the data from your CSV files will be successfully loaded into structured tables within SQL Server Management Studio (SSMS). Specifically, the data will be available in the **Providers** and **PatientAssessments** tables, ready for analysis and reporting.

To view the data you just imported, follow these steps:

- Open SSMS and navigate to the **Providers** and **PatientAssessments** table.
- Right-click on each of the tables.
- Select **"Select Top 1000 Rows"** from the context menu.

This will display the first 1,000 rows of data, allowing you to verify the import and inspect the contents of the tables. You can quickly review your data to confirm that it has been accurately loaded.

```sql
SELECT TOP (1000) [ProviderID]
      ,[ProviderName]
      ,[Phone]
  FROM [Automate_Everything_With_Python].[dbo].[Providers]
```

	ProviderID	ProviderName	Phone
1	6015	Baker, Sara	385-555-0145
2	6350	Zhou, Jon	688-555-0128
3	6541	Carson, Kevin	477-555-0181

```sql
SELECT TOP (1000) [PostPeriod]
      ,[DateEntered]
      ,[ProviderID]
      ,[ClinicID]
      ,[DateTimeAssessment]
      ,[PatientID]
      ,[PatientName]
      ,[Temperature]
      ,[Pulse_Oximeter]
      ,[SOB]
      ,[Cough]
      ,[Abdominal_Pain]
      ,[Diarrhea_Or_Other_Gi_Upset]
      ,[Nausea]
      ,[Loss_Of_Taste]
      ,[Red_Shadowed_Eyes_Or_Pink_Eye]
      ,[Tingling_Sensation_Of_Face_Or_Hands]
      ,[Sore_Throat]
      ,[Chills_And_Or_Repeated_Shaking_With_Chills]
      ,[Muscle_Pain]
      ,[Loss_Of_Smell]
      ,[Headache]
  FROM [Automate_Everything_With_Python].[dbo].[PatientAssessments]
```

	PostPeriod	DateEntered	ProviderID	ClinicID	DateTimeAssessment	PatientID	PatientName
1	202010	2020-10-02 00:00:00.000	6015	79	2020-10-02 12:00:00.000	14020	Kristine, Domingu
2	202010	2020-10-02 00:00:00.000	6015	79	2020-10-02 16:00:00.000	14020	Kristine, Domingu
3	202010	2020-10-01 00:00:00.000	6015	79	2020-10-01 08:00:00.000	14020	Kristine, Domingu

Powerful Data Extraction for Reporting

With this code, you've set up a streamlined workflow that enables data extraction from various files, loading them directly into SSMS. The structured data in **Providers** and **PatientAssessments** is now accessible for generating insightful reports, performing advanced queries, and supporting analytics—all within SSMS, a powerful tool for managing and analyzing data.

- This automation lays the groundwork for efficient and error-free reporting, saving time and making SSMS a central resource for data-driven decision-making.

Tips for Working with Multi-Sheet Excel Files

Here are some best practices when working with multi-sheet Excel files:

- **Flexibility**: The approach used here allows you to handle multiple sheets with different structures from the same Excel file.
- **Data Type Management**: Using converters ensures that data types are consistent between Excel and SQL, preventing data loss or misinterpretation.
- **Efficiency**: The itertuples() method is faster than iterrows(), and using parameterized queries improves the security and efficiency of your SQL commands.

Conclusion

Congratulations! You've successfully learned how to selectively load data from specific Excel worksheets into a SQL database using Python. You now have the skills to:

- Navigate complex Excel workbooks containing multiple sheets.
- Extract data from specific worksheets.
- Insert the relevant data into SQL tables while managing data types effectively.

As you continue to work with Excel and SQL, keep these tips in mind:

- Always inspect your Excel sheets before processing to understand their structure.
- Use the sheet_name parameter to target specific sheets in complex workbooks.
- Be mindful of data types, especially when working with IDs or codes.

With these skills, you're well-prepared to handle even the most complex Excel workbooks and selectively transform their contents into structured SQL databases. Keep practicing, and soon you'll master the process of Excel-to-SQL data integration!

Chapter 4:
From SQL to Excel

Introduction to Automating from SQL to Excel

Welcome back! In this chapter, we'll focus on automating the process of extracting data from a SQL Server database and converting it into formatted Excel reports. This skill is essential for business analysts, data scientists, and managers who need to regularly generate reports from database information. By the end of this chapter, you'll know how to query data from SQL Server, process it with pandas, and create professional Excel reports.

This automated process will not only save time but also reduce the risk of manual errors when generating regular reports.

Code Overview: Automating SQL-to-Excel Transformation

Before diving into the code, let's take a high-level look at the script we'll be building. This script will cover:

- Connecting to SQL Server.
- Querying data and loading it into a pandas DataFrame.
- Processing the data for reporting.
- Exporting the results into an Excel file with formatting.

By the end of this section, you'll be able to automate report generation directly from your SQL database, saving time and effort.

You can find the final script in the **"Python Code"** folder within your **"Automate_Everything_With_Python"** directory for reference. Now, let's start building the script step by step!

```python
# ==============================================================================
# Chapter 4 - From SQL to Excel.py
# ==============================================================================

import os
import pyodbc
import pandas as pd
from openpyxl import load_workbook
from openpyxl.styles import PatternFill
from openpyxl.worksheet.table import Table, TableStyleInfo

# Establish a connection to the SQL Server
connection_string = ('Driver={ODBC Driver 17 for SQL Server};'
                     'Server=Enter_Your_Server_Name;' # Replace with your SQL Server name
                     'Database=Automate_Everything_With_Python;'
                     'UID=automation_user;'
                     'PWD=AutomateEverything2024!;'
                     'Trusted_Connection=no')

conn = pyodbc.connect(connection_string)

# Define the SQL query
sql_query = """
            SELECT
                   p.ProviderID,
                   p.ProviderName,
                   p.Phone,
                   pa.ClinicID,
                   pa.PatientID,
                   pa.PatientName,
                   pa.Temperature,
                   pa.Pulse_Oximeter,
                   pa.DateTimeAssessment
            FROM Automate_Everything_With_Python.dbo.PatientAssessments pa
            LEFT JOIN Automate_Everything_With_Python.dbo.Providers p ON pa.ProviderID = p.ProviderID"""

# Execute the query and fetch the results into a DataFrame
df_patient_readings = pd.read_sql(sql_query, conn)

# Close the connection
conn.close()

# Convert DateTimeAssessment to datetime format
df_patient_readings['DateTimeAssessment'] = pd.to_datetime(df_patient_readings['DateTimeAssessment'])

```

```
# Sort and select the last three records per patient
df_sorted = df_patient_readings.sort_values(by=['PatientID', 'DateTimeAssessment'])
df_last_three = df_sorted.groupby('PatientID').tail(3)
df_last_three.reset_index(drop=True, inplace=True)

# Define paths
desktop_path = os.path.join(os.environ['USERPROFILE'], 'Desktop')
report_folder_path = os.path.join(desktop_path, 'Automate_Everything_With_Python', 'Automated Reports')
os.makedirs(report_folder_path, exist_ok=True)

# Process each provider's data
for provider_id in df_last_three['ProviderID'].unique():
    provider_data = df_last_three[df_last_three['ProviderID'] == provider_id]
    provider_name = provider_data['ProviderName'].values[0]
    excel_file_name = f"{provider_name} - {provider_id} Morning Readings.xlsx"
    excel_file_path = os.path.join(report_folder_path, excel_file_name)
    provider_data.to_excel(excel_file_path, index=False)

    # Load workbook and worksheet
    wb = load_workbook(excel_file_path)
    ws = wb.active

    # Rename worksheet and turn off gridlines
    ws.title = "Data"
    ws.sheet_view.showGridLines = False

    # Create a table in the worksheet
    table_name = "DataTable"
    table = Table(displayName=table_name, ref=ws.dimensions)
    style = TableStyleInfo(name="TableStyleMedium18",
                           showFirstColumn=False,
                           showLastColumn=False,
                           showRowStripes=True,
                           showColumnStripes=False)
    table.tableStyleInfo = style
    ws.add_table(table)

    # Adjust column widths
    for column in ws.columns:
        max_length = max(len(str(cell.value)) for cell in column if cell.value is not None)
        ws.column_dimensions[column[0].column_letter].width = max_length + 2

```

```
    # Apply conditional formatting
    temp_fill = PatternFill(start_color='FFCCCB', end_color='FFCCCB', fill_type='solid')
    oxygen_fill = PatternFill(start_color='83F28F', end_color='83F28F', fill_type='solid')
    for row in range(2, ws.max_row + 1):
        temperature_cell = ws.cell(row=row, column=7)
        pulse_oximeter_cell = ws.cell(row=row, column=8)
        if temperature_cell.value is not None and temperature_cell.value >= 100:
            temperature_cell.fill = temp_fill
        if pulse_oximeter_cell.value is not None and pulse_oximeter_cell.value <= 95:
            pulse_oximeter_cell.fill = oxygen_fill

    # Save the workbook
    wb.save(excel_file_path)

```

Step 1: Importing Libraries

First, let's import the necessary libraries that will help us handle SQL connections, data manipulation, and Excel file creation. Add the following code at the top of your script:

```
import os
import pyodbc
import pandas as pd
from openpyxl import load_workbook
from openpyxl.styles import PatternFill
from openpyxl.worksheet.table import Table, TableStyleInfo
```

Explanation

- **os**: Helps manage file operations.
- **pyodbc**: Allows Python to connect to SQL Server.
- **pandas**: Helps with data manipulation and querying.
- **openpyxl**: Used for handling and formatting Excel files, including reading, writing, and styling.
    - **load_workbook**: Opens and works with existing .xlsx files for reading or modification.
    - **PatternFill**: Provides tools to apply cell background colors, useful for formatting and highlighting data.
    - **Table**: Creates structured tables in Excel worksheets for better organization and usability.
    - **TableStyleInfo**: Applies predefined styles to tables, enhancing their visual appeal and readability.

Step 2: Establishing a SQL Connection

Next, let's define the connection string to connect Python to your SQL Server. This string includes details such as the driver, server name, database name, and login credentials.

```
# Establish a connection to the SQL Server
connection_string = ('Driver={ODBC Driver 17 for SQL Server};'
                     'Server=Enter_Your_Server_Name;' # Replace with your SQL Server name
                     'Database=Automate_Everything_With_Python;'
                     'UID=automation_user;'
                     'PWD=AutomateEverything2024!;'
                     'Trusted_Connection=no')

conn = pyodbc.connect(connection_string)

```

Explanation

- **Driver**: Specifies the ODBC driver to use for the connection.
- **Server**: The name of the SQL Server instance. Replace this with your actual server information.
- **Database**: The name of the database to which you're connecting.
- **UID**: This stands for "User ID." In Python, the `UID` is used to specify the username that will be used to authenticate the connection to the SQL Server. It represents the account under which the connection is made.
- **PWD**: This stands for "Password." The `PWD` is the corresponding password for the username (UID) that grants access to the SQL Server. In Python, this is used along with the `UID` to authenticate the connection and ensure that only authorized users can access the database.
- **Trusted_Connection=no**: Specifies whether to use Windows authentication. If you set this to yes, the connection will use your Windows credentials, and you can omit the UID and PWD.
- **pyodbc.connect()**: This function establishes a connection to a database using the **ODBC** (Open Database Connectivity) driver. It requires a **connection string** that contains details from the connection string as the driver type, server, database, username, password, and any other necessary parameters to connect to the database.
- **connection_string**: This variable contains all the necessary information for connecting to the SQL Server database, including the driver to use, the server address, database name, and credentials (UID and PWD).
- **conn**: This variable represents the connection object that is created once the `connect()` function is called successfully. The connection object allows you to interact with the SQL Server database, such as executing SQL queries, retrieving data, and managing transactions.

Step 3: Querying Data from SQL Server

Now that we're connected to the SQL Server, let's define the SQL query and retrieve the data into a pandas DataFrame:

```
# Define the SQL query
sql_query = """
            SELECT
                p.ProviderID,
                p.ProviderName,
                p.Phone,
                pa.ClinicID,
                pa.PatientID,
                pa.PatientName,
                pa.Temperature,
                pa.Pulse_Oximeter,
                pa.DateTimeAssessment
            FROM Automate_Everything_With_Python.dbo.PatientAssessments pa
            LEFT JOIN Automate_Everything_With_Python.dbo.Providers p ON pa.ProviderID = p.ProviderID"""

```

Explanation

- **SELECT**: This clause specifies the columns you want to retrieve from the database. In this case, you are selecting various columns from two different tables: PatientAssessments (aliased as pa) and Providers (aliased as p).

  The selected columns include:

  - **p.ProviderID**: The ID of the provider, coming from the Providers table.
  - **p.ProviderName**: The name of the provider from the Providers table.
  - **p.Phone:** The phone number of the provider.
  - **pa.ClinicID**: The ID of the clinic where the patient was assessed, from the PatientAssessments table.
  - **pa.PatientID**: The unique ID of the patient.
  - **pa.PatientName**: The name of the patient.
  - **pa.Temperature**: The patient's temperature recorded during the assessment.
  - **pa.Pulse_Oximeter**: The patient's pulse oximeter reading (oxygen saturation).
  - **pa.DateTimeAssessment**: The date and time when the patient was assessed.
- **FROM Automate_Everything_With_Python.dbo.PatientAssessments pa**: This specifies the table from which the main data is being selected. In this case, you are selecting from the PatientAssessments table in the Automate_Everything_With_Python database, and the table is aliased as pa for shorter reference.
- **LEFT JOIN Automate_Everything_With_Python.dbo.Providers p ON pa.ProviderID = p.ProviderID**:
  This performs a **LEFT JOIN** between the PatientAssessments table and the Providers table. A **LEFT JOIN** returns all rows from the PatientAssessments table, even if there are no matching rows in the Providers table.
  - The join condition ON pa.ProviderID = p.ProviderID ensures that the data from Providers is included only when the ProviderID in PatientAssessments matches a ProviderID in Providers. If no matching provider exists for a given patient assessment, the provider-related columns (p.ProviderID, p.ProviderName, etc.) will return NULL.

```
# Execute the query and fetch the results into a DataFrame
df_patient_readings = pd.read_sql(sql_query, conn)

```

Explanation

- **pd.read_sql()**: This function from pandas is used to execute the SQL query on the SQL Server database and load the results directly into a pandas DataFrame. A **DataFrame** is a powerful and flexible data structure in pandas that functions like a table, similar to an Excel spreadsheet, where data can be easily manipulated and analyzed.
- **sql_query**: This is the SQL query you wrote earlier. It fetches data from the PatientAssessments and Providers tables based on the selection and joins specified.
- **conn**: This is the connection object created by pyodbc.connect(). It is passed to pd.read_sql() so that the function knows which database to execute the query on.

Once the query is executed, the results are stored in the **DataFrame** called df_patient_readings. Each column in the SQL result corresponds to a column in the DataFrame. You can now use pandas functions to analyze and manipulate the data in df_patient_readings.

Closing the Connection

```
# Close the connection
conn.close()
```

- **conn.close()**: This method closes the connection to the SQL Server once you're done fetching data. It's important to close the connection to free up resources and avoid locking the database or running into connection limits.

Summary

1. **SQL Query**: The query selects data from two tables (PatientAssessments and Providers) and performs a LEFT JOIN to include information about the provider for each patient assessment.
2. **Data Fetching**: pd.read_sql() is used to execute the SQL query and retrieve the data into a pandas DataFrame called df_patient_readings. This DataFrame can be used to analyze and manipulate the data in Python.
3. **Closing the Connection**: Once the data has been retrieved, the connection to the database is closed using conn.close() to ensure that resources are released.

Step 4: Processing the Data

Once we have the data, let's process it to get it ready for reporting:

Converting the 'DateTimeAssessment' Column to a Datetime Format

```
# Convert DateTimeAssessment to datetime format
df_patient_readings['DateTimeAssessment'] = pd.to_datetime(df_patient_readings['DateTimeAssessment'])
```

Explanation

- **df_patient_readings**: This is a pandas DataFrame that contains the data fetched from the SQL query. One of the columns in this DataFrame is 'DateTimeAssessment', which stores the date and time when a patient's assessment was recorded.
- **pd.to_datetime()**: This pandas function converts the values in the 'DateTimeAssessment' column from strings (or other formats) into proper datetime objects. This is critical because:

    - **Why Convert to Datetime?**: Datetime objects allow you to perform various date-related operations, such as sorting, filtering by date, and performing calculations (e.g., finding the time difference between two events).
    - Without this conversion, pandas would treat the date values as strings, making it difficult to perform meaningful operations (e.g., sorting by date, comparing dates).

In essence, this step ensures that 'DateTimeAssessment' is stored in a format that allows for easy manipulation and analysis of the date and time data.

```
# Sort and select the last three records per patient
df_sorted = df_patient_readings.sort_values(by=['PatientID', 'DateTimeAssessment'])
df_last_three = df_sorted.groupby('PatientID').tail(3)
df_last_three.reset_index(drop=True, inplace=True)
```

Explanation

- **sort_values()**: This function sorts the rows in the DataFrame based on the values in one or more columns.
- **by=['PatientID', 'DateTimeAssessment']**: This tells pandas to sort the DataFrame first by 'PatientID' (which groups all the assessments belonging to each patient together), and within each patient's group, it further sorts the assessments by 'DateTimeAssessment' in ascending order (from earliest to latest).
    - **Why Sort?**: Sorting by 'PatientID' and 'DateTimeAssessment' ensures that each patient's data is organized chronologically. This is particularly important when you want to focus on the most recent records for each patient (as is done in the next step).
-
- **groupby('PatientID')**: This groups the sorted DataFrame by 'PatientID', meaning all records related to the same patient are considered together as a group.
    - **Why Grouping Matters?**: Grouping allows you to perform operations on each group individually. In this case, we want to isolate the records for each patient so that we can focus on selecting the last three records for each group (i.e., each patient).
- **tail(3)**: After grouping the data by patient, this function selects the latest three records for each group (in this case, for each patient). These three records/rows correspond to the most recent assessments, thanks to the earlier sorting step.
    - **Why Use tail(3)?**: The goal is to retrieve the last three assessments for each patient. Since the data is already sorted chronologically, tail(3) will select the three most recent assessments for every patient.

- **reset_index()**: This function resets the index of the DataFrame. When pandas performs operations like grouping, the index can become disordered or misaligned with the new structure of the DataFrame. Resetting the index restores it to a clean, sequential order.

- **drop=True**: By setting drop=True, the old index is discarded entirely rather than being added as a new column in the DataFrame.
- **inplace=True**: This modifies the DataFrame in place, meaning the change happens directly in df_last_three without needing to assign the result to a new variable.

Why Reset the Index?

When you manipulate data using functions like groupby or tail(), the original row indices (the row numbers from the original DataFrame) can be retained. This can cause confusion if you plan to further manipulate or analyze the data. Resetting the index ensures that the DataFrame has a fresh, clean index starting from 0, making it easier to work with in subsequent operations.

Summary

1. **Convert 'DateTimeAssessment' to Datetime**: This ensures that the assessment times are treated as actual datetime objects, which allows for sorting and date-related calculations.
2. **Sort by 'PatientID' and 'DateTimeAssessment'**: The data is first grouped by each patient, and then sorted chronologically within each group. This prepares the data for selecting the most recent records.
3. **Select the Last Three Records per Patient**: By using groupby and tail(3), you are selecting the three most recent assessment records for each patient.
4. **Reset the Index**: This step ensures that the final DataFrame, df_last_three, has a clean and sequential index, which makes further operations easier to perform.

Why Is This Important?

This process is particularly useful when you are analyzing longitudinal data—data that tracks changes over time for the same individual. By selecting the most recent records for each patient, you can focus your analysis on current or recent data, which is often critical in healthcare, business, and time-sensitive domains.

Step 5: Setting Up File Paths

Next, let's define the folder paths where our Excel reports will be saved:

Building and Creating a Folder for Reports

```
# Define paths
desktop_path = os.path.join(os.environ['USERPROFILE'], 'Desktop')
report_folder_path = os.path.join(desktop_path, 'Automate_Everything_With_Python', 'Automated Reports')
os.makedirs(report_folder_path, exist_ok=True)

```

This code is responsible for constructing a dynamic path to a folder on the user's desktop where reports will be saved and ensuring that the folder exists. Let's break it down step by step.

1. Constructing the Desktop Path

- **os.path.join()**: This function from the os module is used to construct file or directory paths. Instead of manually typing out paths, which can vary between operating systems (Windows, macOS, Linux), os.path.join() automatically handles the correct path separators (\ on Windows, / on macOS/Linux) to ensure compatibility.
    - In this case, it takes two arguments:
        - **os.environ['USERPROFILE']**: This accesses the USERPROFILE environment variable, which stores the path to the current user's profile directory (e.g., C:\Users\YourUsername\ on Windows). This ensures that the script works regardless of who is logged in to the machine.
        - **'Desktop'**: This string is appended to the user's profile path, resulting in the path to the user's desktop (e.g., C:\Users\YourUsername\Desktop).
- **Why Use os.path.join()?**: It avoids hardcoding file paths, ensuring that the script will work on any machine, regardless of where the user's profile or desktop is located. It also ensures compatibility across different operating systems by automatically handling the correct file path format.

  Result: The desktop_path variable now contains the full path to the current user's desktop (e.g., C:\Users\YourUsername\Desktop).

2. Constructing the Report Folder Path

- **os.path.join()**: Once again, os.path.join() is used to build a dynamic path to the folder where the reports will be stored. This time, it appends two additional folder names to the desktop path:
    - **'Automate_Everything_With_Python'**: This is the name of a parent folder where all related files for this project will be stored.
    - **'Automated Reports'**: This is the subfolder inside the project folder where the automated reports will be saved.
- **Why Structure the Path This Way?**: Organizing files into specific folders helps keep the desktop clean and maintains a clear structure for your project. By creating a project folder (Automate_Everything_With_Python) and a dedicated folder for reports (Automated Reports), you ensure that all generated reports are kept in one place, making them easier to find and manage.

  Result: The report_folder_path variable now contains the full path to the directory where the reports will be saved (e.g., C:\Users\YourUsername\Desktop\Automate_Everything_With_Python\Automated Reports).

3. Creating the Folder if It Doesn't Exist

- **os.makedirs()**: This function is responsible for creating directories. It works like the command mkdir but can also create parent directories if they don't exist. In this case, it attempts to create the full path specified in report_folder_path.

- **exist_ok=True**: This parameter ensures that the function does not raise an error if the folder already exists. Without this parameter, trying to create a folder that already exists would result in a FileExistsError. By setting exist_ok=True, the code will silently continue if the folder is already present.
- **Why Use os.makedirs()?**: This function ensures that the necessary folder structure is in place to save the reports. Even if the folders don't exist yet, it will create them, ensuring that the script always has a place to store the generated Excel files.

  Result: If the 'Automate_Everything_With_Python' and 'Automated Reports' folders don't already exist on the desktop, they will be created. If they do exist, the code continues without any errors.

Summary

1. **os.path.join()**: Dynamically builds the full path to the folder where reports will be stored, ensuring compatibility across operating systems and avoiding hardcoding paths. This makes the code more flexible and reusable.
2. **os.makedirs()**: Ensures that the folder exists, creating it if necessary. This guarantees that the folder structure is always ready to store the generated reports, even if the script is run on a new machine or after deleting the folders.

By using these functions, you automate the process of ensuring that your reports are stored in the correct location, without worrying about whether the folders exist or whether the code will work across different machines. This approach keeps your project well-organized and portable.

Step 6: Generating Excel Reports

Now let's generate Excel reports for each provider using the data we processed earlier:

Processing Data and Creating Excel Reports for Each Provider

This section of the code loops through each provider's data, exports it to an Excel file, and formats the Excel sheet with custom styling and conditional formatting.

```
# Process each provider's data
for provider_id in df_last_three['ProviderID'].unique():
    provider_data = df_last_three[df_last_three['ProviderID'] == provider_id]
    provider_name = provider_data['ProviderName'].values[0]
    excel_file_name = f"{provider_name} - {provider_id} Morning Readings.xlsx"
    excel_file_path = os.path.join(report_folder_path, excel_file_name)
    provider_data.to_excel(excel_file_path, index=False)

```

Explanation

- **df_last_three['ProviderID'].unique()**: This part identifies the unique Provider IDs in the df_last_three DataFrame. The .unique() function ensures that the loop runs once for each provider, avoiding duplicate reports.
- **provider_data = df_last_three[df_last_three['ProviderID'] == provider_id]**: For each unique provider, this line filters the DataFrame df_last_three to get all the data associated with that particular provider. It creates a smaller DataFrame, provider_data, containing only the relevant rows for that provider.
- **provider_name = provider_data['ProviderName'].values[0]**: This extracts the provider's name from the first row of the provider_data DataFrame. Since all rows for a given provider have the same name, you only need to retrieve the name from the first row.
- **excel_file_name**: This string uses Python's f-string syntax to dynamically name the Excel file based on the provider's name and ID. For example, the file might be named "Baker, Sara - 6015 Morning Readings.xlsx".
- **os.path.join()**: Combines the report folder path and the dynamically created file name to generate the full file path where the report will be saved.
- **to_excel()**: This pandas function exports the provider_data DataFrame to an Excel file at the specified path (excel_file_path). The index=False argument ensures that the DataFrame's index (row numbers) is not written to the Excel file, keeping the output clean and focused on the data itself.

Loading the Excel File and Accessing the Worksheet

```
    # Load workbook and worksheet
    wb = load_workbook(excel_file_path)
    ws = wb.active

```

Explanation

- **load_workbook()**: This function from the openpyxl library loads the Excel file (excel_file_path) that was just created. This allows you to modify the file further by adding tables, adjusting formatting, and applying styles.
- **wb.active**: This retrieves the currently active worksheet (the one where the data was written). By default, the active worksheet is the first worksheet in the workbook.

Renaming the Worksheet and Turning Off Gridlines

```
    # Rename worksheet and turn off gridlines
    ws.title = "Data"
    ws.sheet_view.showGridLines = False

```

Explanation

- **ws.title**: Renames the worksheet to "Data," which gives the worksheet a more meaningful name.
- **ws.sheet_view.showGridLines = False**: Turns off the gridlines in the worksheet, which can make the Excel report look cleaner and more professional.

Creating a Table in the Worksheet

```python
    # Create a table in the worksheet
    table_name = "DataTable"
    table = Table(displayName=table_name, ref=ws.dimensions)
    style = TableStyleInfo(name="TableStyleMedium18",
                           showFirstColumn=False,
                           showLastColumn=False,
                           showRowStripes=True,
                           showColumnStripes=False)
    table.tableStyleInfo = style
    ws.add_table(table)

```

Explanation

- **Table(displayName="DataTable", ref=ws.dimensions)**: This creates a new table in the worksheet using the entire range of data (ws.dimensions gives the boundaries of the data). The displayName="DataTable" assigns a name to the table.
- **TableStyleInfo()**: This defines the table's style:
  - **name="TableStyleMedium18"**: Specifies the predefined style to apply. In this case, the style is "TableStyleMedium18," which includes alternating row stripes and other formatting.
  - **showRowStripes=True**: Enables row stripes for better readability.
- **ws.add_table(table)**: Adds the styled table to the worksheet.

Adjusting Column Widths

```python
    # Adjust column widths
    for column in ws.columns:
        max_length = max(len(str(cell.value)) for cell in column if cell.value is not None)
        ws.column_dimensions[column[0].column_letter].width = max_length + 2

```

Explanation

- **for column in ws.columns**: Loops through each column in the worksheet.
- **max(len(str(cell.value)) for cell in column if cell.value is not None)**: Calculates the maximum length of the content in each column. It converts each cell's value to a string (in case of numeric data) and checks the length, ignoring any empty cells (if cell.value is not None).

- **ws.column_dimensions[column[0].column_letter].width = max_length + 2**: Adjusts the column width based on the maximum content length, with an extra 2 units of padding for readability.

Applying Conditional Formatting

```
# Apply conditional formatting
temp_fill = PatternFill(start_color='FFCCCB', end_color='FFCCCB', fill_type='solid')
oxygen_fill = PatternFill(start_color='83F28F', end_color='83F28F', fill_type='solid')
for row in range(2, ws.max_row + 1):
    temperature_cell = ws.cell(row=row, column=7)
    pulse_oximeter_cell = ws.cell(row=row, column=8)
    if temperature_cell.value is not None and temperature_cell.value >= 100:
        temperature_cell.fill = temp_fill
    if pulse_oximeter_cell.value is not None and pulse_oximeter_cell.value <= 95:
        pulse_oximeter_cell.fill = oxygen_fill

```

Explanation

- **PatternFill()**: This is part of the openpyxl library and is used to define a fill color for a cell:
    - **temp_fill**: A solid red fill is applied for cells that meet the condition for high temperature (>= 100°F).
    - **oxygen_fill**: A green fill is applied for cells with low pulse oximeter readings (≤ 95%).
- **for row in range(2, ws.max_row + 1)**: This loop goes through each row in the worksheet, starting from the second row (the first row is typically the header row).
- **temperature_cell = ws.cell(row=row, column=7)**: Retrieves the cell in column 7 (which corresponds to the 'Temperature' column) for each row.
- **pulse_oximeter_cell = ws.cell(row=row, column=8)**: Retrieves the cell in column 8 (which corresponds to the 'Pulse Oximeter' column) for each row.
- **Conditional Formatting**:
    - If the temperature is 100°F or higher, the temperature cell gets the red fill (temp_fill).
    - If the pulse oximeter reading is 95% or lower, the pulse oximeter cell gets the green fill (oxygen_fill).

Saving the Workbook

```
# Save the workbook
wb.save(excel_file_path)

```

Explanation

- **wb.save()**: Saves the Excel workbook with the updates, including the new table, column width adjustments, and conditional formatting.

Summary

1. **Loop Through Providers**: The code loops through each provider's data, creating a separate Excel file for each provider.
2. **Export to Excel**: It exports the provider's data to an Excel file using pandas.
3. **Format the Excel File**: The script then loads the Excel file and applies custom formatting:
    - Renames the worksheet and removes gridlines.
    - Adds a formatted table for the data.
    - Adjusts column widths to fit the content.
    - Applies conditional formatting to highlight specific values (high temperatures and low pulse oximeter readings).
4. **Save the File**: Finally, the script saves the formatted Excel report.

This step-by-step process ensures that each provider's data is exported, formatted, and saved in a professional-looking Excel report.

Final Output

After running the code, you'll now have a set of formatted Excel reports generated directly from your SQL Server data. Here's what the output includes:

1. **Excel Reports for Each Provider**: In the folder path you specified, you'll find individual Excel files for each provider, named according to the provider's name and ID. Each report includes:
    - A structured table of patient assessment data.
    - Automatically adjusted column widths for clear readability.
    - Conditional formatting that highlights critical values, such as high temperatures and low pulse oximeter readings, making key information stand out.

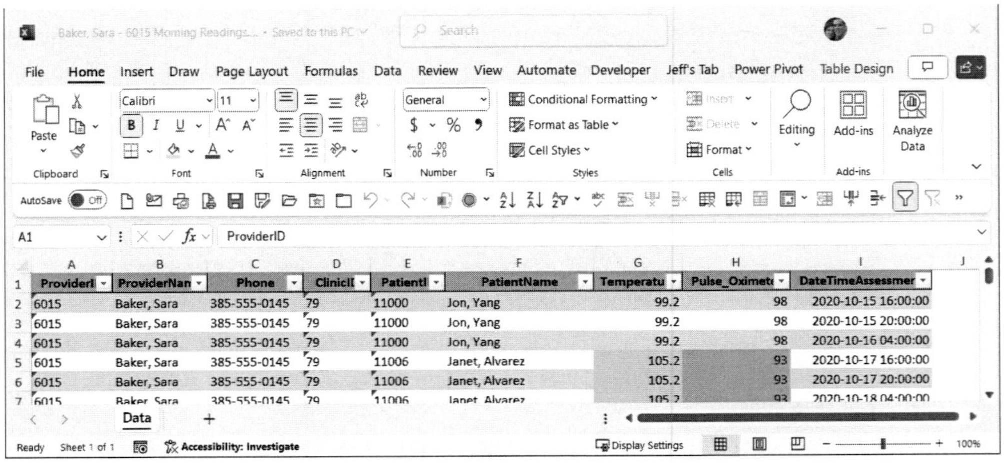

2. **Organized Folder Structure**: These reports are neatly saved in the **"Automated Reports"** folder inside your **"Automate_Everything_With_Python"** directory on your desktop. This organized structure ensures easy access and keeps all generated reports in one convenient location.

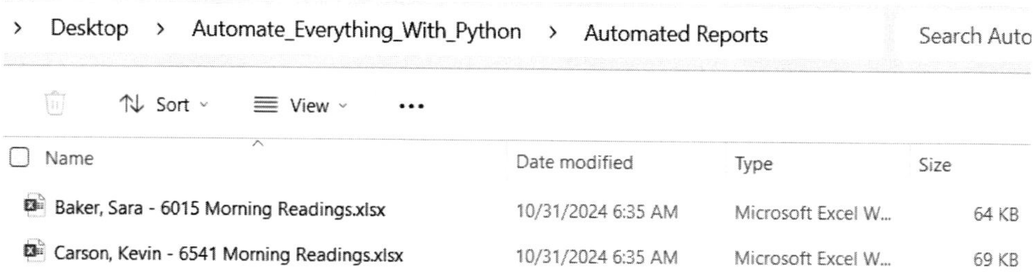

3. **SQL Data Transformed for Reporting**: The data from the **PatientAssessments** and **Providers** tables has been extracted, processed, and structured within each report. You can verify that the information has been formatted and grouped for easy readability, allowing you to quickly identify trends and insights.
4. **Efficient, Reusable Workflow**: The code you just executed is now a powerful, reusable tool for generating similar reports from SQL data. Need to run the report again? Simply execute the code to refresh the data or adjust the SQL query to tailor the reports for other datasets.

Benefits of Your Automated Report Workflow

By completing this exercise, you've created an automated reporting system that:

- **Saves Time**: Generate reports instantly without the need for manual exporting or formatting.
- **Reduces Errors**: Automated data handling minimizes the risk of manual errors, ensuring consistent and accurate reports.
- **Supports Flexible Reporting**: You can modify this script to suit various reporting needs, making it a versatile tool in your data automation toolkit.

You now have a fully automated SQL-to-Excel report generator at your fingertips, empowering you to produce professional reports with minimal effort. This foundation will be invaluable as you continue to expand your skills in Python and automation!

Conclusion

In this chapter, you have successfully learned how to automate the process of generating professional Excel reports directly from your SQL Server data. By building a flexible and efficient Python script, you've gained the ability to:

- **Query Data from SQL Server**: Using pandas, you've seen how to extract data from your database and load it into a DataFrame for easy manipulation.
- **Process and Organize Data for Reporting**: With tools like sorting, grouping, and filtering, you can now prepare your data efficiently before exporting it.
- **Generate Excel Reports with Advanced Formatting**: You've learned to create well-structured and styled Excel files using openpyxl, including:
    - Adding tables to worksheets.
    - Adjusting column widths to fit the content.
    - Applying conditional formatting to highlight critical data like high temperatures and low pulse oximeter readings.

This script serves as the foundation for building a robust reporting system that eliminates the need for manual work and reduces the potential for errors. However, this is only the beginning. Here are some ways you can expand on what you've learned:

- **Add More Complex Data Transformations**: Modify your script to handle more advanced calculations, aggregate data, or add custom formulas in Excel.
- **Create Charts and Pivot Tables**: Incorporate visual elements like charts and pivot tables to enhance your reports and make data insights clearer.
- **Add Error Handling and Logging**: Make your script more resilient by adding mechanisms to catch errors, log events, and notify users of any issues during the report generation process.
- **Automate Report Distribution**: Take it to the next level by automating the distribution of reports via email or uploading them to shared drives.

By mastering these techniques, you'll be equipped to handle increasingly complex data and reporting needs. Keep practicing and experimenting, and soon you'll be able to build sophisticated, automated reporting workflows that save time and improve accuracy across your organization.

Chapter 5: Automating Python Workflows

Introduction to Python Workflow Automation

Welcome back! In this chapter, we will explore the powerful technique of automating Python workflows by executing Python scripts programmatically. This method allows you to orchestrate complex workflows, schedule tasks, and manage dependencies between scripts, making it highly valuable for data engineers, DevOps professionals, and anyone who needs to automate multi-step processes.

By the end of this chapter, you'll know how to:

- Execute other Python scripts programmatically.
- Capture and handle their output.
- Gracefully manage errors that might occur during execution.

Let's get started and automate your Python workflows with confidence!

Code Overview: Automating Python Script Execution

Before diving into the code, here's a high-level overview of what we'll accomplish:

- **Execute another Python script** using a Python program.
- **Capture the output** of the executed script.
- **Handle errors** and unexpected issues gracefully.

This script will automate the execution of other Python scripts in a controlled manner. By the end of this section, you'll be able to integrate it into your workflow automation systems.

You can find the final script in the **"Python Code"** folder within your **"Automate_Everything_With_Python"** directory for reference. Let's break the process down step by step.

```python
# ==============================================================================
# Chapter 5 - Automating Python Workflows.py
# ==============================================================================

import subprocess
import sys
import os

# Define paths
desktop_path = os.path.join(os.environ['USERPROFILE'], 'Desktop')
report_folder_path = os.path.join(desktop_path, 'Automate_Everything_With_Python', 'Python Code')

# Name of the script to execute
script_name = "Chapter 4 - From SQL to Excel.py"
script_to_execute = os.path.join(report_folder_path, script_name)

try:
    # Execute the script using subprocess
    result = subprocess.run([sys.executable, script_to_execute], capture_output=True, text=True, check=True)

    # Print the output
    print(f"Script '{script_name}' executed successfully.")
    print("Output:")
    print(result.stdout)

except subprocess.CalledProcessError as e:
    print(f"An error occurred while executing the script: {e}")
    print("Error output:")
    print(e.stderr)
except FileNotFoundError:
    print(f"The script '{script_to_execute}' was not found.")
except Exception as e:
    print(f"An unexpected error occurred: {e}")

```

Step 1: Importing Libraries and Setting Up File Paths

First, let's import the necessary libraries and set up the file paths that point to the Python script we want to execute. Add the following code to the top of your script:

```python
import subprocess
import sys
import os

# Define paths
desktop_path = os.path.join(os.environ['USERPROFILE'], 'Desktop')
report_folder_path = os.path.join(desktop_path, 'Automate_Everything_With_Python', 'Python Code')

# Name of the script to execute
script_name = "Chapter 4 - From SQL to Excel.py"
script_to_execute = os.path.join(report_folder_path, script_name)
```

Explanation

- **subprocess**: This module allows Python to spawn new processes, connect to input/output/error pipes, and obtain return codes from other commands. It's essential for executing external programs, including other Python scripts.
- **sys**: Used here to access variables related to the Python interpreter, such as sys.executable, which ensures the correct interpreter is used to run the script.
- **os**: Provides functions to interact with the operating system, especially for file path manipulation.
- **desktop_path**: This is the path to the current user's desktop, dynamically built using the USERPROFILE environment variable. Using os.path.join() ensures compatibility across different operating systems.
- **report_folder_path**: The folder where our Python scripts are stored. We are dynamically constructing the path to the "Python Code" folder on the desktop.
- **script_name**: The name of the Python script we want to execute.
- **script_to_execute**: Combines the report_folder_path and the script_name to form the full path to the Python script we intend to run.

Summary

In this step, we imported the necessary libraries and dynamically built the file path to the Python script we want to execute. This ensures that the path will work on any system where the code is run, regardless of the operating system or user profile.

Step 2: Executing the Script and Handling Output

Now, let's write the core functionality that executes the Python script and handles its output:

```
try:
    # Execute the script using subprocess
    result = subprocess.run([sys.executable, script_to_execute], capture_output=True, text=True, check=True)

    # Print the output
    print(f"Script '{script_name}' executed successfully.")
    print("Output:")
    print(result.stdout)

except subprocess.CalledProcessError as e:
    print(f"An error occurred while executing the script: {e}")
    print("Error output:")
    print(e.stderr)
except FileNotFoundError:
    print(f"The script '{script_to_execute}' was not found.")
except Exception as e:
    print(f"An unexpected error occurred: {e}")

```

Explanation

- **try-except block**: This structure is used to handle potential errors that could occur during the execution of the script. Let's break down each part:
- **subprocess.run()**:
  - **[sys.executable, script_to_execute]**: Executes the specified Python script using the same Python interpreter that is running this script (sys.executable ensures the correct Python version is used).
  - **capture_output=True**: Captures the output (both stdout and stderr) of the executed script.
  - **text=True**: Ensures the output is returned as a string, making it easier to process and display.
  - **check=True**: Raises a CalledProcessError if the script returns a non-zero exit code, which is commonly used to indicate errors.
- **result.stdout**: This prints the standard output from the executed script if it completes successfully.
- **except subprocess.CalledProcessError**: This exception is raised if the script returns a non-zero exit code. In this block, we print an error message along with the captured error output (e.stderr).
- **except FileNotFoundError**: Handles the case where the specified Python script is not found in the expected location.
- **except Exception as e**: A generic exception handler that catches any other errors that might occur during the script execution and prints an informative error message.

Summary

In this step, we executed the target Python script using the subprocess module, captured its output, and handled any potential errors with appropriate messages. This ensures that the script's behavior is tracked and if any issues are reported to the user.

Final Output

After executing the code in this chapter, you will see updated Excel reports generated with a new timestamp, reflecting the most recent date and time the files were created. Here's what this process accomplished:

1. **Re-running the Code from Chapter 4**: This code has triggered the **"Chapter 4 - From SQL to Excel.py"** Python script, which automates the entire workflow from data extraction to report generation. By automating the execution of a previous Python script within this new script, you've streamlined the process, allowing for easy updates to your reports without manual re-running of each script.
2. **Automatic Execution of VBA Code**: Within the Chapter 4 Python script, VBA macros embedded in the Excel files were also executed, updating the files with any necessary

changes or formatting applied by the VBA code. This seamless integration of Python and VBA ensures that the final Excel files are both updated and professionally formatted without any additional manual intervention.

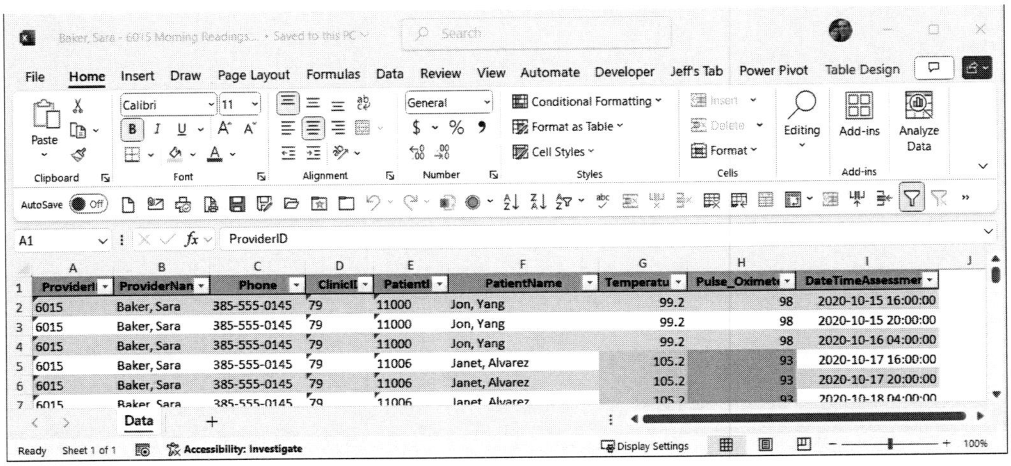

3. **Updated Files with Timestamps**: You can confirm that the reports were recreated by checking the date and time stamp on each Excel file. The timestamp will indicate when the file was last generated, allowing you to verify that the latest data and formatting are in place.

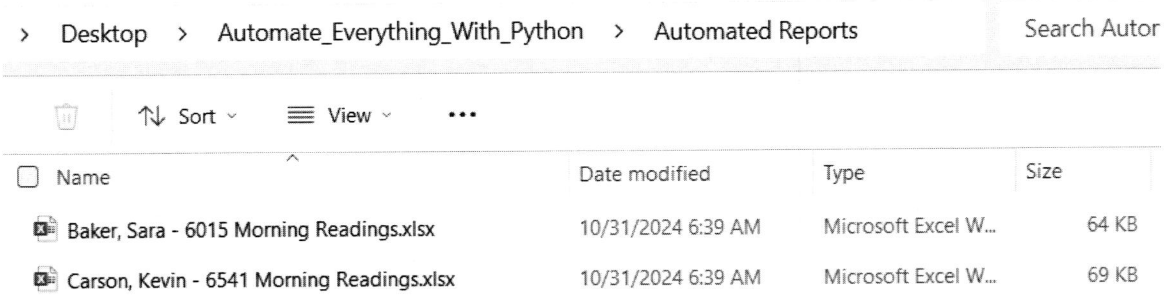

4. **Executing Python Code with Python Code**: This chapter demonstrated how you can use Python to call and run other Python scripts. By organizing code in this way, you've created a flexible, layered automation system that allows you to run entire workflows with a single command, making updates and maintenance significantly more efficient.

Benefits of This Automated Workflow

By automating the execution of previous scripts, you gain:

- **Efficiency**: No need to open and run each individual script; the workflow handles everything in one execution.

- **Consistency**: Ensures that all data is refreshed, and formatting is applied in the same way each time the code runs.
- **Scalability**: This modular approach makes it easy to add or update scripts in the workflow, allowing you to expand functionality over time.

With this setup, you can recreate Excel reports anytime by simply running this chapter's code, confident that it will execute the entire workflow from data extraction to report formatting automatically. This powerful technique of executing code with Python code enhances flexibility and control over your data automation projects.

Conclusion

Congratulations! You've just learned how to automate the execution of Python scripts, manage their output, and handle potential errors. Let's recap the powerful skills you've gained:

- **Using the subprocess module** to execute other Python scripts programmatically.
- **Capturing and displaying the output** of these scripts to monitor their behavior.
- **Handling various types of errors**, including missing files, script execution failures, and unexpected exceptions.
- **Using environment variables and os.path** to dynamically manage file paths in a cross-platform manner.

Your script is now a powerful tool that can:

- Execute other Python scripts programmatically.
- Capture and display the output of these scripts.
- Handle errors gracefully, providing informative error messages.

Expanding Your Automation

As you continue to work with Python automation, consider enhancing this script by:

- **Executing multiple scripts in sequence**: Extend the script to run several Python scripts in a defined order.
- **Implementing logging**: Keep track of when and how scripts were executed by adding logging functionality.
- **Adding command-line arguments**: Allow the script to dynamically accept different scripts or parameters via command-line inputs for greater flexibility.
- **Integrating with a task scheduler**: Automate the execution of this script at regular intervals using a task scheduler like cron (Linux) or Task Scheduler (Windows).

For even more advanced automation, you might explore:

- **Using configuration files**: Manage script parameters using configuration files like YAML or JSON.
- **Parallel execution**: Run multiple Python scripts in parallel to speed up execution.
- **Integrating with cloud services**: Use cloud services like AWS Lambda or Azure Functions to manage and execute distributed workflows.

By mastering these techniques, you'll be able to automate increasingly complex workflows and streamline processes within your organization.

Chapter 6: Automating Monthly Reporting Workflows

Introduction to Monthly Report Automation

Welcome back! In this chapter, we will learn how to automate the organization and archiving of monthly reports using Python. This workflow will create folders for future months, copy reports into those folders, and append date stamps to the filenames. This technique is highly valuable for businesses that generate recurring reports and need an efficient way to maintain an organized file structure.

By the end of this chapter, you'll have a Python script that automates the process of managing recurring reports, saving time and ensuring consistency.

Code Overview: Automating Monthly Report Organization

Before diving into the code, here's an overview of what the script will accomplish:

- Create folders for the next 12 months.
- Copy existing reports into those folders.
- Add date stamps to filenames to enhance tracking.

The final code will be stored in the **"Python Code"** folder within your **"Automate_Everything_With_Python"** directory for easy reference. Now, let's get started with building your monthly report automation!

```python
# ============================================================================
# Chapter 6 - Automating Monthly Reporting Workflows.py
# ============================================================================

import os
from datetime import datetime
from dateutil.relativedelta import relativedelta
import shutil

# Define the base path for the report folders
desktop_path = os.path.join(os.environ['USERPROFILE'], 'Desktop')
report_folder_path = os.path.join(desktop_path, 'Automate_Everything_With_Python', 'Automated Reports')

# Define the source path where the Excel files are located
source_path = report_folder_path  # Assuming the Excel files are on the desktop

# Get the current date
current_date = datetime.now()

```

```python
    # Create monthly folders for the next 12 months
    for i in range(12):
        # Calculate the folder name in the format 'yyyy-mm Reports'
        folder_name = (current_date + relativedelta(months=i)).strftime('%Y-%m Reports')

        # Create the full path for the folder
        folder_path = os.path.join(report_folder_path, folder_name)

        # Create the folder if it doesn't exist
        if not os.path.exists(folder_path):
            os.makedirs(folder_path)
            print(f"Created folder: {folder_path}")
        else:
            print(f"Folder already exists: {folder_path}")

        # Find and copy any files ending with 'Morning Readings.xlsx'
        for file_name in os.listdir(source_path):
            if file_name.endswith("Morning Readings.xlsx"):
                # Construct the source file path
                source_file_path = os.path.join(source_path, file_name)

                # Create the new file name with the datestamp
                datestamp = current_date.strftime('%Y-%m-%d')
                new_file_name = f"{file_name[:-5]} {datestamp}.xlsx"

                # Construct the destination file path
                destination_file_path = os.path.join(folder_path, new_file_name)

                # Copy the file to the destination folder
                shutil.copy2(source_file_path, destination_file_path)
                print(f"Copied {file_name} to {destination_file_path}")

```

Step 1: Importing Libraries and Setting Up File Paths

First, we need to import the necessary libraries and define the file paths that the script will use. Add this code at the top of your script:

```python
import os
from datetime import datetime
from dateutil.relativedelta import relativedelta
import shutil

# Define the base path for the report folders
desktop_path = os.path.join(os.environ['USERPROFILE'], 'Desktop')
report_folder_path = os.path.join(desktop_path, 'Automate_Everything_With_Python', 'Automated Reports')

# Define the source path where the Excel files are located
source_path = report_folder_path  # Assuming the Excel files are on the desktop

```

Explanation

- **os**: This module provides functions to interact with the operating system, allowing you to work with file and directory paths across different platforms. This ensures compatibility whether the script is run on Windows, macOS, or Linux.
- **datetime**: This module provides classes for manipulating dates and times. Here, we use it to get the current date and time, which will later be used for folder names and date stamps.
- **relativedelta**: A function from the dateutil library that simplifies date arithmetic. For example, it makes it easy to compute future or past dates by months, days, or years.
- **shutil**: A high-level module that provides utilities for file operations such as copying and moving files. In this script, it's used to copy files from one folder to another while preserving metadata such as timestamps.
- **desktop_path**: This variable constructs the path to the user's desktop. By using os.environ['USERPROFILE'], we dynamically retrieve the profile path of the current user.
- **report_folder_path**: This variable defines the path to the "Automated Reports" folder on the user's desktop. This is where the monthly report folders will be created.
- **source_path**: This defines where the existing reports (Excel files) are located. For this script, we assume that the source files are in the same "Automated Reports" folder, but this can be customized depending on your file structure.

Summary

In this step, we imported the necessary libraries and constructed dynamic paths to ensure our script works on any system. The paths point to the user's desktop and the report storage folder. We also set up the source path where the existing reports are located.

Step 2: Getting the Current Date

Next, we need to capture the current date, which will be used for folder names and date stamps:

```
# Get the current date
current_date = datetime.now()

```

Explanation

- **datetime.now()**: This method retrieves the current date and time from the system clock. In this script, it is used to append date stamps to filenames and create folder names based on the current month and year.

- **Why Use the Current Date?**: By using the current date, we can dynamically name folders and files without manually specifying the date each time the script is run. This helps keep the workflow consistent, automated, and flexible.

Summary

We retrieved the current date and time, which will be used for naming folders and appending date stamps to the reports.

Step 3: Creating Monthly Folders and Copying Reports

Now, let's write the core functionality that creates folders for the next 12 months and copies reports into them with date stamps:

```
# Create monthly folders for the next 12 months
for i in range(12):
    # Calculate the folder name in the format 'yyyy-mm Reports'
    folder_name = (current_date + relativedelta(months=i)).strftime('%Y-%m Reports')

    # Create the full path for the folder
    folder_path = os.path.join(report_folder_path, folder_name)

    # Create the folder if it doesn't exist
    if not os.path.exists(folder_path):
        os.makedirs(folder_path)
        print(f"Created folder: {folder_path}")
    else:
        print(f"Folder already exists: {folder_path}")

    # Find and copy any files ending with 'Morning Readings.xlsx'
    for file_name in os.listdir(source_path):
        if file_name.endswith("Morning Readings.xlsx"):
            # Construct the source file path
            source_file_path = os.path.join(source_path, file_name)

            # Create the new file name with the datestamp
            datestamp = current_date.strftime('%Y-%m-%d')
            new_file_name = f"{file_name[:-5]} {datestamp}.xlsx"

            # Construct the destination file path
            destination_file_path = os.path.join(folder_path, new_file_name)

            # Copy the file to the destination folder
            shutil.copy2(source_file_path, destination_file_path)
            print(f"Copied {file_name} to {destination_file_path}")
```

Explanation

- **for i in range(12)**: This loop iterates 12 times, once for each month, starting from the current month.
- **relativedelta(months=i)**: This adds i months to the current date (current_date), ensuring that the script generates folders for the upcoming months. For example, if i=1, the folder will be created for the next month.
- **strftime('%Y-%m Reports')**: This formats the folder name in the yyyy-mm format followed by "Reports." For example, the folder for November 2024 would be named 2024-11 Reports.
- **if not os.path.exists(filepath)**: Returns True if the file or directory specified by filepath does not exists, otherwise it returns False
- **os.makedirs()**: This method creates a new folder in the specified path (folder_path). If the if not os.path.exists(filepath) returns False indicating a folder already exists, it is skipped.
- **os.listdir(source_path)**: This function lists all the files in the source_path. The script looks for files that end with 'Morning Readings.xlsx' and processes them.
- **shutil.copy2()**: This function copies the files from the source directory to the destination folder. The copy2() method is used instead of copy() to ensure that file metadata, such as the original creation date and time, is preserved.
- **datestamp**: This variable holds the current date in the yyyy-mm-dd format and is used to create the new filename with the appended date stamp. For example, if the file was originally named Clinic Morning Readings.xlsx, the new file would be named Clinic Morning Readings 2024-11-01.xlsx.
- **Print Statements**: These print statements serve as logs to notify the user about which folders are created and which files are copied. This is helpful for tracking the script's progress.

Summary

In this step, we created folders for the next 12 months, searched for existing report files in the source directory, and copied them into the corresponding monthly folder. The script also appended a date stamp to each file's name for better organization and tracking.

Final Output

After executing the code, you'll see a well-organized folder structure on your desktop under the **"Automate Everything With Python\Automated Reports"** directory. Here's what the code accomplished:

1. **Monthly Folders Created**: The script generated folders for each of the next 12 months, with names like **"2024-11 Reports"**, **"2024-12 Reports"**, and so on. This consistent naming convention makes it easy for departments to locate monthly reports and ensures that folders remain in chronological order.

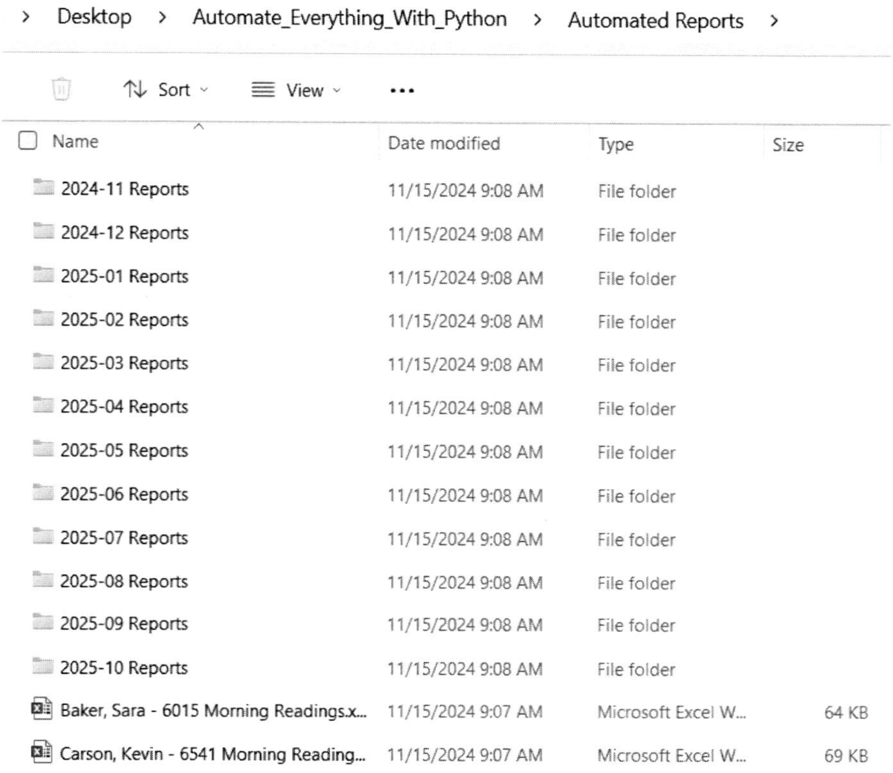

2. **Date-Stamped Reports in Each Folder**: Existing report files have been copied into each monthly folder with a date stamp added to the filename. For example, a report originally named **"Clinic Morning Readings.xlsx"** will now appear as **"Clinic Morning Readings 2024-11-01.xlsx"**. This date-stamped naming convention is beneficial for tracking when each report was generated, aiding in file versioning and historical reference.

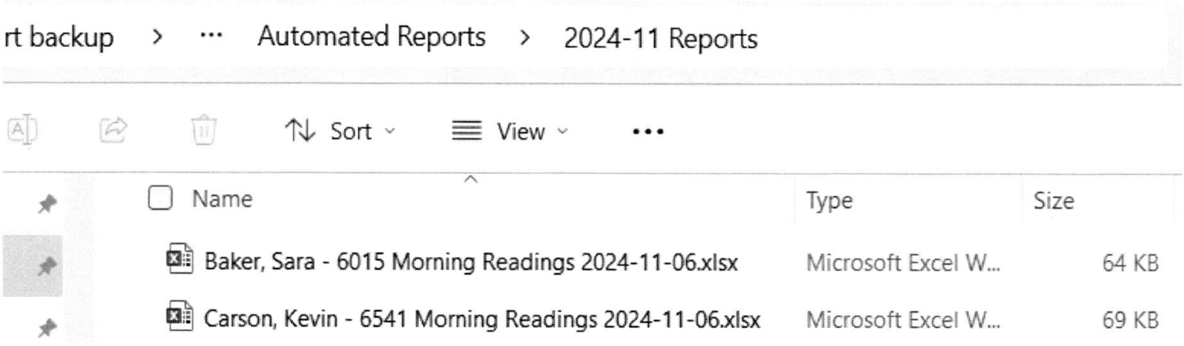

3. **Efficient Report Organization for Departments**: With this automated folder structure, departments can easily access monthly reports without the risk of overwriting older files.

The standardized naming convention ensures that all files are neatly organized and ready for use in recurring monthly reports, reducing manual work and minimizing errors.

Why This Automation Is Valuable

This automated structure not only saves time by creating folders and copying files systematically, but it also ensures that reports are archived in a clear, predictable format. This method is especially useful for departments that need to generate and access monthly reports regularly. By following a consistent structure, the script provides a scalable solution that can be used month after month with minimal maintenance.

Conclusion

Congratulations! You've successfully automated your monthly reporting workflow. Here's a recap of the skills you've acquired in this chapter:

- **Using the os module** for file and directory operations.
- **Working with dates** and performing date arithmetic using datetime and relativedelta.
- **Copying files** and preserving metadata with shutil.
- **Creating an organized folder structure** for managing reports across multiple months.
- **Adding date stamps to filenames** for better tracking and version control.

With this script, you can now:

- Automatically create a folder structure for the next 12 months.
- Copy and organize recurring reports into their respective folders.
- Append date stamps to filenames to ensure that you can track and manage multiple versions of your reports.

You're now equipped with a powerful tool to streamline your monthly reporting workflows. Keep refining your skills, and you'll continue to find ways to optimize

Chapter 7: Automating Personalized Monthly Report Distribution

Introduction to Automated Report Distribution

In this chapter, we'll create a Python script that automates the distribution of personalized monthly reports via email. This technique is invaluable for businesses that need to send individualized reports to multiple recipients on a regular basis. By the end of this chapter, you will know how to read recipient information from an Excel file, match it with the correct report, and send personalized emails with attachments using Outlook.

This skill is especially valuable for data analysts, HR professionals, and anyone managing recurring report distribution tasks. Let's dive in and automate your report distribution process!

Before We Start: Update the Email Table

Please ensure you have entered a valid email address in the **"Email Table.xlsx"** workbook. This file is located in the **"Data Files"** folder within the **"Automate_Everything_With_Python"** directory on your desktop.

To receive a test email when the code executes, follow these steps:

1. **Open the "Email Table.xlsx" Workbook**: Navigate to the **"Data Files"** folder and open the **"Email Table.xlsx"** file.

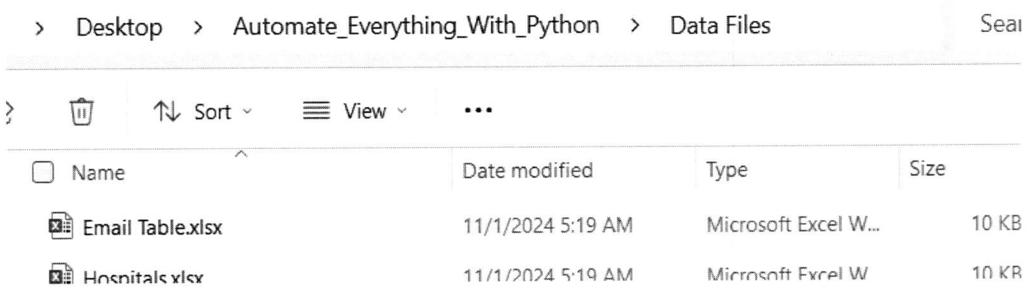

2. **Enter Your Email Address**: In the **"Email Address"** column, add a valid email address, preferably your own. This will ensure that you receive a sample email when the code runs.

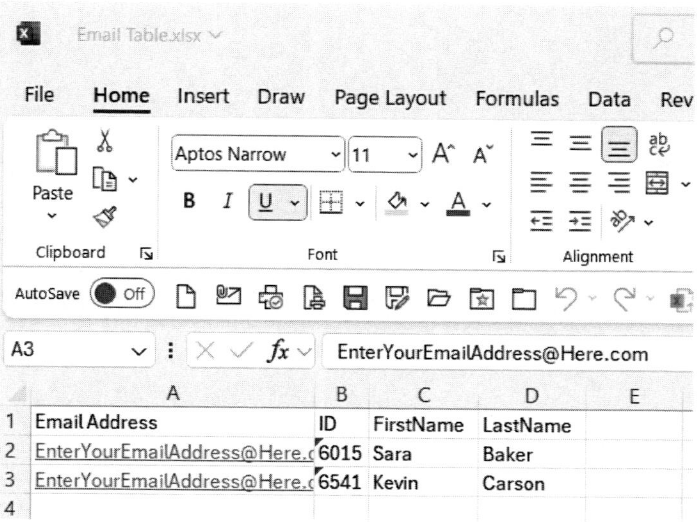

3. **Save and Close the Workbook**: Once you've entered your email address, save the workbook and close it.

This updated file will be used throughout the process, so ensure the email address is correct and accessible for testing. Once you've completed this step, you're ready to move on!

Code Overview: Automating Report Distribution

Before we jump into the details, here's an overview of what our script will achieve:

- **Load recipient information** from an Excel file.
- **Match each recipient** with their respective report.
- **Send personalized emails** with the report as an attachment using Outlook.

The final code will be stored in the **"Python Code"** folder within your **"Automate_Everything_With_Python"** directory. Now, let's walk through the steps to build your report distribution automation.

```python
import os
import pandas as pd
import win32com.client as win32
from datetime import datetime

# Define paths
desktop_path = os.path.join(os.environ['USERPROFILE'], 'Desktop')
data_files_path = os.path.join(desktop_path, 'Automate Everything With Python', 'Data Files')
email_table_path = os.path.join(data_files_path, 'Email Table.xlsx')
report_folder_path = os.path.join(desktop_path, 'Automate Everything With Python', 'Automated Reports')

# Load the "Email Table.xlsx" into a DataFrame
df_email = pd.read_excel(email_table_path)

# Get the current month's folder path
current_month_folder_name = datetime.now().strftime('%Y-%m Reports')
current_month_folder_path = os.path.join(report_folder_path, current_month_folder_name)

# Initialize Outlook application
outlook = win32.Dispatch('outlook.application')
```

```python

# Process each Excel file in the current month's folder
for file_name in os.listdir(current_month_folder_path):
    if file_name.endswith('.xlsx'):
        # Extract the 4-digit ID from the file name
        try:
            id_number = file_name.split('-')[1].strip().split()[0]
        except IndexError:
            print(f"Skipping file: {file_name}, unable to extract ID.")
            continue

        # Find the corresponding email address in the DataFrame
        email_row = df_email[df_email['ID'].astype(str) == id_number]
        if not email_row.empty:
            recipient_email = email_row['Email Address'].values[0]
            first_name = email_row['FirstName'].values[0]
            last_name = email_row['LastName'].values[0]

            # Create a new email
            mail = outlook.CreateItem(0)
            mail.Subject = f"Monthly Report - {current_month_folder_name}"
            mail.Body = f"Please find the attached report for: {first_name} {last_name}."
            mail.To = recipient_email

            # Attach the file to the email
            file_path = os.path.join(current_month_folder_path, file_name)
            mail.Attachments.Add(file_path)
            print(f"Attached file: {file_name} for ID: {id_number} to {recipient_email}")

            # Send the email
            mail.Send()
            print(f"Email sent to {recipient_email} for file: {file_name}")
        else:
            print(f"No email found for ID: {id_number}, skipping file: {file_name}")

```

Step 1: Importing Libraries and Setting Up File Paths

We start by importing the necessary libraries and defining the file paths that point to the email table and the report folders. Add the following code at the top of your script:

```
import os
import pandas as pd
import win32com.client as win32
from datetime import datetime

# Define paths
desktop_path = os.path.join(os.environ['USERPROFILE'], 'Desktop')
data_files_path = os.path.join(desktop_path, 'Automate_Everything_With_Python', 'Data Files')
email_table_path = os.path.join(data_files_path, 'Email Table.xlsx')
report_folder_path = os.path.join(desktop_path, 'Automate_Everything_With_Python', 'Automated Reports')
```

Explanation

- **os**: Handles file and directory operations, allowing us to work across different platforms (Windows, macOS, Linux).
- **pandas (pd)**: A powerful library for data manipulation. We use it to load and process the Excel file containing recipient data.
- **win32com.client (win32)**: This module connects Python to Outlook, enabling us to automate the sending of emails.
- **datetime**: Allows us to work with dates. We use it to create the current month's folder name.
- **desktop_path**: This variable dynamically retrieves the path to the user's desktop using the USERPROFILE environment variable.
- **data_files_path**: This is where the "Email Table.xlsx" file is stored, containing recipient information.
- **email_table_path**: This is the full path to the Excel file that holds the recipient data (names, email addresses, and IDs).
- **report_folder_path**: This is the path to the folder where monthly reports are stored.

Summary

We imported the necessary libraries and set up dynamic paths to the email table and report folders. These paths ensure the script can access the required data, regardless of the system it's run on.

Step 2: Loading Email Information and Determining the Current Month's Folder

Next, we load the recipient information from the "Email Table.xlsx" file and determine the folder for the current month's reports.

```
# Load the "Email Table.xlsx" into a DataFrame
df_email = pd.read_excel(email_table_path)

# Get the current month's folder path
current_month_folder_name = datetime.now().strftime('%Y-%m Reports')
current_month_folder_path = os.path.join(report_folder_path, current_month_folder_name)
```

Explanation

- **pd.read_excel()**: This function loads the Excel file (email table) into a pandas DataFrame, making it easy to manipulate and filter the data.
- **datetime.now().strftime('%Y-%m Reports')**: This line retrieves the current date and formats it as yyyy-mm Reports, which will match the name of the folder containing the monthly reports.
- **os.path.join()**: Dynamically constructs the full path to the current month's report folder.

Summary

We loaded recipient information into a pandas DataFrame and generated the folder path for the current month's reports based on the current date. This ensures the script always processes the correct reports for each month.

Step 3: Initializing the Outlook Application

Now, we set up our connection to Outlook, which we'll use to send the personalized emails.

```
# Initialize Outlook application
outlook = win32.Dispatch('outlook.application')

```

Explanation

- **win32.Dispatch('outlook.application')**: This command establishes a connection to the Outlook application. It allows the script to create, send, and manage emails directly through Outlook.

Summary

We initialized a connection to Outlook, which will be used to send the personalized emails with the reports attached.

Step 4: Processing and Sending Reports

This step contains the core logic that processes the reports and sends personalized emails to each recipient.

```python
        # Find the corresponding email address in the DataFrame
        email_row = df_email[df_email['ID'].astype(str) == id_number]
        if not email_row.empty:
            recipient_email = email_row['Email Address'].values[0]
            first_name = email_row['FirstName'].values[0]
            last_name = email_row['LastName'].values[0]

            # Create a new email
            mail = outlook.CreateItem(0)
            mail.Subject = f"Monthly Report - {current_month_folder_name}"
            mail.Body = f"Please find the attached report for: {first_name} {last_name}."
            mail.To = recipient_email

            # Attach the file to the email
            file_path = os.path.join(current_month_folder_path, file_name)
            mail.Attachments.Add(file_path)
            print(f"Attached file: {file_name} for ID: {id_number} to {recipient_email}")

            # Send the email
            mail.Display()
            # mail.Send()
            print(f"Email sent to {recipient_email} for file: {file_name}")
        else:
            print(f"No email found for ID: {id_number}, skipping file: {file_name}")
```

Explanation

- **os.listdir()**: Lists all files in the current month's folder. The script processes only files that end with '.xlsx' (Excel files).
- **file_name.split()**: Extracts the 4-digit ID from the filename, which is used to match the report with the correct recipient.
- **df_email[df_email['ID'].astype(str) == id_number]**: Filters the DataFrame to find the recipient whose ID matches the one extracted from the filename.
- **outlook.CreateItem(0)**: Creates a new email item in Outlook.
- **mail.Subject**: Sets the subject of the email to include the current month's folder name.
- **mail.Body**: Customizes the email body with the recipient's first and last name.
- **mail.Attachments.Add()**: Attaches the corresponding report to the email.
- **mail.Send()**: Sends the email to the recipient.

Summary

In this step, the script processes each report, extracts the recipient's ID from the filename, looks up the corresponding email address in the DataFrame, and sends a personalized email with the report attached. It prints status messages for each action to track progress.

Final Output

After executing this script, personalized monthly reports will be automatically sent via email to recipients based on the information in the **"Email Table.xlsx"** file. Here's a breakdown of the process and final results:

1. **Loading Email and Recipient Information**: The script loads the **"Email Table.xlsx"** file into a DataFrame. This file contains recipient information, including email addresses, first names, last names, and unique IDs. The IDs will be used to match each report to the correct recipient.
2. **Identifying the Current Month's Reports**: The script dynamically identifies the folder for the current month (e.g., **"2024-10 Reports"**) within the **"Automated Reports"** directory on your desktop. It processes each Excel file in this folder, looking for files that need to be emailed.
3. **Matching Report Files to Recipients**: For each Excel file in the folder, the script attempts to extract a 4-digit ID from the file name. This ID is used to look up the corresponding recipient's information in the DataFrame. If a matching ID is found, the script retrieves the recipient's email address, first name, and last name.
4. **Creating and Sending Personalized Emails**: Using the retrieved information, the script creates a personalized email for each recipient:
   - The email's subject includes the current month's report period (e.g., **"Monthly Report - 2024-10 Reports"**).
   - The email body includes a personalized message with the recipient's full name.
   - The report file is attached to the email.
5. **Automatic Report Distribution**: Once each email is prepared, the script sends it automatically through Outlook, attaching the corresponding report file. For each successful email, the script outputs a message confirming the file attachment and email delivery. If no matching email is found for a report's ID, the script skips that file and prints a notification.

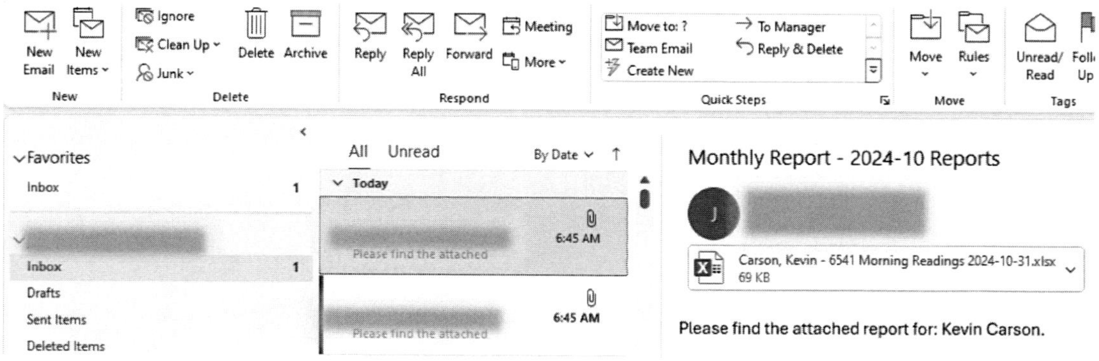

Summary

By executing this code, you have:

- **Automated the Monthly Report Distribution**: No manual emails or file attachments are required; the script handles the entire process.
- **Ensured Accuracy in Report Delivery**: Each report is sent to the correct recipient based on their unique ID, minimizing the risk of errors.
- **Personalized Emails for Professionalism**: Each recipient receives a tailored email, making the communication more professional and enhancing the recipient experience.

This automated solution saves time, reduces manual effort, and ensures consistent, timely delivery of personalized monthly reports, making it an efficient tool for managing recurring report distribution.

Conclusion

Congratulations! You've successfully automated the process of distributing personalized monthly reports. Let's recap the powerful skills you've learned:

- **Using pandas** to read and process data from Excel files.
- **Navigating file systems** to locate and process reports.
- **Automating Outlook** to send personalized emails with attachments.
- **Handling exceptions** and providing informative output for troubleshooting.

Your script is now a powerful tool that can:

- Personalize each email based on recipient information.
- Match report files with recipient data, ensuring each person receives their specific report.
- Save significant time and reduce errors in the report distribution process.

Expanding the Script

As you continue automating report distribution, consider enhancing the script by:

- **Adding error logging** to track issues during email distribution.
- **Implementing retry logic** for failed email sends.
- **Creating a summary report** of successful and failed distributions.

With these skills, you're now equipped to streamline your report distribution workflow and improve communication efficiency across your organization!

Chapter 8:
Automated Backup Solutions

Introduction to Automated Backups

In this chapter, we'll dive into a critical aspect of managing your Python projects automating the backup of your Python code files. Automating backups is essential for developers and data scientists who want to maintain regular, organized backups of their work. By the end of this chapter, you'll know how to create date-stamped backup folders and copy Python files into them automatically.

This skill is particularly valuable for version control, disaster recovery, and maintaining a history of your code changes. Let's start building your automated backup solution and ensure your projects are always safe!

Code Overview: Automating Backup Creation

Before diving into the code, here's a high-level overview of what the script will do:

- **Create date-stamped folders** to organize backups.
- **Copy Python code files** into the backup folder.
- **Display informative messages** to track the progress of the backup.

This backup solution will automate the process of creating and managing backups for your Python code files. You can find the final script in the **"Python Code"** folder within your **"Automate_Everything_With_Python"** directory. Let's walk through the steps to create your backup solution.

```python
import os
import shutil
from datetime import datetime

# Define paths
desktop_path = os.path.join(os.environ['USERPROFILE'], 'Desktop')
python_code_path = os.path.join(desktop_path, 'Automate Everything With Python', 'Python Code')
backup_folder_path = os.path.join(desktop_path, 'Automate Everything With Python','Python Backup Files')

# Create the backup folder if it doesn't exist
if not os.path.exists(backup_folder_path):
    os.makedirs(backup_folder_path)

# Create a subfolder with the current date
current_date = datetime.now().strftime('%Y-%m-%d')
date_folder_path = os.path.join(backup_folder_path, current_date)

if not os.path.exists(date_folder_path):
    os.makedirs(date_folder_path)

# Copy all Python files to the date folder
for file_name in os.listdir(python_code_path):
    if file_name.endswith('.py'):  # Check for Python files
        src_file_path = os.path.join(python_code_path, file_name)
        dest_file_path = os.path.join(date_folder_path, file_name)
        shutil.copy(src_file_path, dest_file_path)
        print(f"Copied: {file_name} to {date_folder_path}")

print(f"Backup completed successfully in {date_folder_path}.")
```

Step 1: Importing Libraries and Setting Up File Paths

We begin by importing the necessary libraries and setting up paths for the backup folder and the files to be backed up:

```python
import os
import shutil
from datetime import datetime

# Define paths
desktop_path = os.path.join(os.environ['USERPROFILE'], 'Desktop')
python_code_path = os.path.join(desktop_path, 'Automate_Everything_With_Python', 'Python Code')
backup_folder_path = os.path.join(desktop_path, 'Automate_Everything_With_Python','Python Backup Files')

```

Explanation

- **os**: This module provides functions to interact with the file system, enabling us to create directories, list files, and manage paths.
- **shutil**: This high-level module provides functions for copying files and directories, making it perfect for our backup solution.
- **datetime**: This module is used to capture the current date, which will be used for creating date-stamped backup folders.

- **desktop_path**: Dynamically retrieves the user's desktop path using the USERPROFILE environment variable.
- **python_code_path**: This path points to the folder containing Python files that need to be backed up.
- **backup_folder_path**: This path points to the main backup folder where all backups will be stored.

Summary

We've imported the necessary libraries and set up dynamic paths for both the Python files to be backed up and the location of the backup folder. This ensures our script works consistently across different systems and environments.

Step 2: Creating the Main Backup Folder

Next, we need to create the main backup folder if it doesn't already exist:

```
# Create the backup folder if it doesn't exist
if not os.path.exists(backup_folder_path):
    os.makedirs(backup_folder_path)
```

Explanation

- **os.path.exists()**: Checks if the backup folder already exists. If the folder doesn't exist, it returns False.
- **os.makedirs()**: Creates the main backup folder, allowing us to store all backups in a structured location.

Summary

In this step, we ensured that a backup folder exists by checking if it's already there and creating it if needed. This guarantees that all backups will have a designated location.

Step 3: Creating a Date-Stamped Subfolder for Backups

We now create a date-stamped subfolder to organize each backup by the date it was created:

```python
# Create a subfolder with the current date
current_date = datetime.now().strftime('%Y-%m-%d')
date_folder_path = os.path.join(backup_folder_path, current_date)

if not os.path.exists(date_folder_path):
    os.makedirs(date_folder_path)
```

Explanation

- **datetime.now().strftime('%Y-%m-%d')**: This generates the current date in the format YYYY-MM-DD, which will be used as the folder name for the backup.
- **date_folder_path**: This variable constructs the full path for the date-stamped subfolder inside the main backup folder.
- **os.makedirs()**: Creates the date-stamped subfolder if it doesn't already exist, organizing each backup by the date it was performed.

Summary

We created a date-stamped subfolder inside the main backup folder to keep backups organized by the date they were created. This structure makes it easier to track and manage multiple backups over time.

Step 4: Copying Python Files to the Backup Folder

Now we write the core functionality of the script, which copies all Python files from the source folder to the backup folder:

```python
# Copy all Python files to the date folder
for file_name in os.listdir(python_code_path):
    if file_name.endswith('.py'):  # Check for Python files
        src_file_path = os.path.join(python_code_path, file_name)
        dest_file_path = os.path.join(date_folder_path, file_name)
        shutil.copy(src_file_path, dest_file_path)
        print(f"Copied: {file_name} to {date_folder_path}")

print(f"Backup completed successfully in {date_folder_path}.")
```

Explanation

- **os.listdir()**: Lists all files in the Python code folder (python_code_path), allowing the script to identify the files that need to be backed up.
- **file_name.endswith('.py')**: This checks whether the file is a Python file (i.e., it ends with the .py extension). Only Python files will be copied.
- **shutil.copy()**: Copies the Python file from the source folder (python_code_path) to the destination folder (date_folder_path). This function also preserves file metadata, such as the creation and modification timestamps.
- **print()**: Outputs a message for each file that is successfully copied, providing feedback on the progress of the backup.

Summary

In this step, we copied all Python files from the source folder to the date-stamped backup folder. The script outputs progress messages for each file that is copied and prints a final message once the backup is complete. This ensures that the backup process is transparent and traceable.

Final Output

After running this code, a backup of all Python files in your **"Python Code"** folder will be created. Here's a step-by-step breakdown of what the script does:

1. **Backup Folder Setup**: The script first checks if a **"Python Backup Files"** folder exists within your **"Automate_Everything_With_Python"** directory on the desktop. If it doesn't exist, the script creates it, ensuring there's always a designated space for backups.
2. **Date-Stamped Subfolder Creation**: Within the **"Python Backup Files"** folder, the script creates a new subfolder named with the current date in **YYYY-MM-DD** format (e.g., **2024-11-15**). This ensures that each backup is saved in a uniquely dated folder, making it easy to organize and retrieve backups by the date they were created.

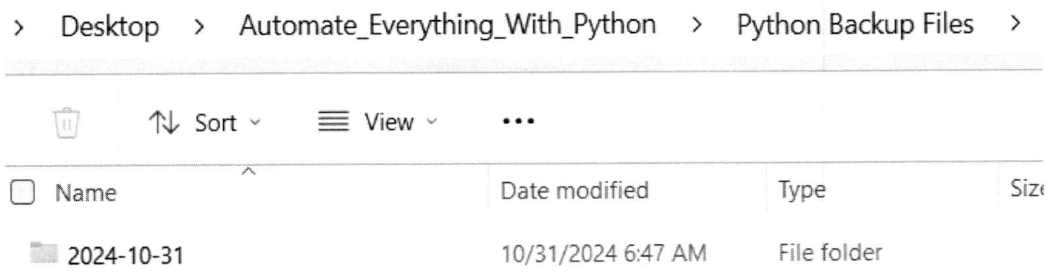

3. **Copying Python Files**: The script scans the **"Python Code"** folder for any files ending with **.py** (Python scripts). It copies each Python file into the date-stamped backup folder.

For each file copied, the script outputs a message confirming the file transfer, allowing you to track which files were backed up.

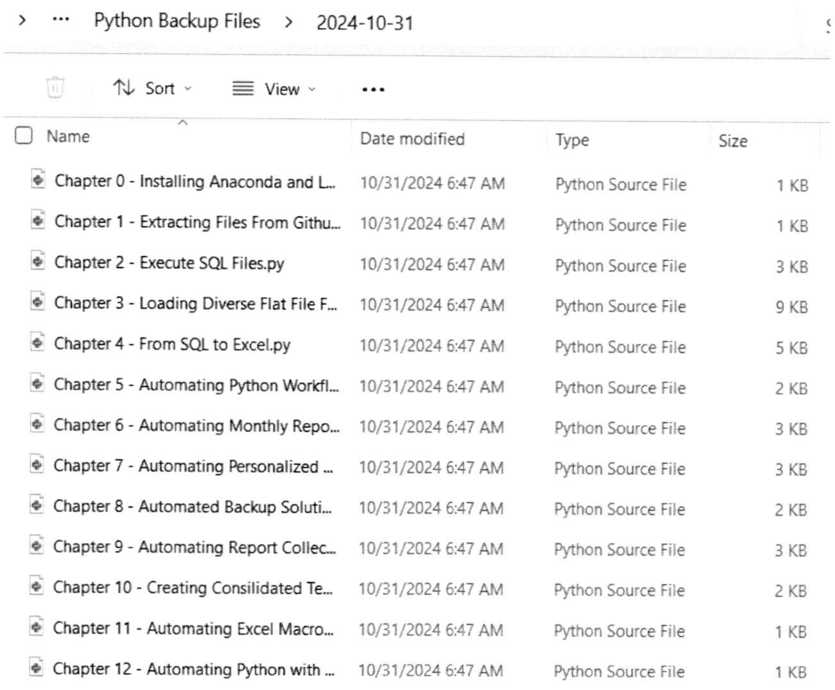

4. **Completion Message**: Once all Python files are copied, a final message is printed, confirming that the backup was completed successfully and indicating the location of the backup.

Benefits of This Automated Backup Solution

By running this code, you have:

- **Ensured Data Security**: All Python scripts are backed up with a date stamp, providing a safety net in case of accidental deletion or file corruption.
- **Organized Backup System**: The date-stamped folders make it easy to locate backups based on when they were created, simplifying file recovery.
- **Scalable and Reusable Workflow**: You can run this code anytime to create an updated backup, ensuring that you always have the latest versions of your Python files safely stored.

This automated backup solution offers peace of mind and easy access to previous versions of your scripts, supporting a well-organized and secure file management system.

Conclusion

Congratulations! You've successfully built an automated backup solution for your Python code. Here's a recap of the skills you've acquired in this chapter:

- **Using os and shutil modules** for file and directory operations.
- **Creating date-stamped folders** to organize backups by date.
- **Copying Python files** based on their file extension.
- **Providing progress feedback** during the backup process.

Your script is now a powerful tool that can:

- Automatically create date-stamped folders for easy tracking of backup history.
- Copy Python files to the backup folder and preserve metadata.
- Provide clear output of the backup process, ensuring transparency.

Taking it Further

As you continue to automate your Python workflows, consider enhancing this backup script by:

- **Implementing incremental backups** that only copy changed or new files.
- **Adding error handling and logging** for better reliability and troubleshooting.
- **Integrating with cloud storage services** like Google Drive or AWS for off-site backups.

For more advanced backup automation, you could explore:

- **Creating a user-friendly interface** for managing backup settings and scheduling.
- **Adding a scheduler** to run backups automatically at specific intervals (e.g., daily, weekly).
- **Integrating with version control** systems for more robust backup and versioning.

With this automated backup script, you're now well-equipped to safeguard your Python projects and maintain a clear history of code versions. Keep refining and expanding your skills, and your backup solutions will become even more powerful!

Chapter 9: Automating Report Collection and Aggregation from Outlook

Introduction to Automated Report Collection and Aggregation

In this chapter, we will automate the process of collecting reports from Outlook emails and aggregating them into a single Excel file. This is a crucial workflow for managers and data analysts who regularly receive reports via email and need a streamlined way to combine them into one comprehensive report. By the end of this chapter, you'll be able to interact with Outlook, download email attachments, and seamlessly combine multiple Excel files into a single report.

This automation will save you countless hours of manual work, freeing up time for higher-value tasks. Let's dive in!

Code Overview: Automating Report Collection

Here's an overview of what the script will do:

- **Connect to Outlook** and access the inbox.
- **Download attachments** from relevant emails.
- **Load and combine Excel files** into one consolidated report.

This script will automate the entire process of collecting reports from emails and aggregating them, ensuring you never have to do it manually again. The final code is saved in the **"Python Code"** folder within the **"Automate_Everything_With_Python"** directory. Now, let's walk through each step of building this powerful automation.

```python
# ============================================================================
# Chapter 9 - Automating Report Collection and Aggregation From Outlook.py
# ============================================================================

import os
import pandas as pd
import win32com.client as win32

# Define the folder path for downloaded emails
desktop_path = os.path.join(os.environ['USERPROFILE'], 'Desktop')
downloaded_folder_path = os.path.join(desktop_path, 'Automate Everything With Python', 'Email Downloaded Folder')

# Create the downloaded folder if it doesn't exist
if not os.path.exists(downloaded_folder_path):
    os.makedirs(downloaded_folder_path)

# Initialize Outlook application
outlook = win32.Dispatch('outlook.application')
namespace = outlook.GetNamespace('MAPI')

# Access the inbox
inbox = namespace.GetDefaultFolder(6)  # 6 refers to the inbox folder
messages = inbox.Items

# Initialize a list to store DataFrames
dataframes = []

# Process each email in the inbox
for message in messages:
    if message.Subject.startswith("Monthly Report - ") and message.Attachments.Count > 0:
        # Download each attachment in the email
        for attachment in message.Attachments:
            file_name = attachment.FileName
            save_path = os.path.join(downloaded_folder_path, file_name)
            attachment.SaveAsFile(save_path)
            print(f"Downloaded: {file_name} to {downloaded_folder_path}")

            # Check if the file is an Excel file
            if file_name.endswith('.xlsx'):
                # Load the Excel file into a DataFrame
                df = pd.read_excel(save_path)
                dataframes.append(df)

```

```python
# Concatenate all DataFrames into a single DataFrame
if dataframes:
    combined_df = pd.concat(dataframes, ignore_index=True)
    combined_df_path = os.path.join(downloaded_folder_path, 'Combined_Report.xlsx')
    combined_df.to_excel(combined_df_path, index=False)
    print(f"Combined DataFrame saved to: {combined_df_path}")
else:
    print("No relevant emails with attachments found.")

```

Step 1: Importing Libraries and Setting Up File Paths

First, we need to import the necessary libraries and define the paths where the email attachments will be saved:

```python
import os
import pandas as pd
import win32com.client as win32

# Define the folder path for downloaded emails
desktop_path = os.path.join(os.environ['USERPROFILE'], 'Desktop')
downloaded_folder_path = os.path.join(desktop_path, 'Automate_Everything_With_Python', 'Email Downloaded Folder')

# Create the downloaded folder if it doesn't exist
if not os.path.exists(downloaded_folder_path):
    os.makedirs(downloaded_folder_path)
```

Explanation

- **os**: This module handles file and directory operations, allowing us to create folders and manage file paths.
- **pandas (pd)**: This library is used for reading and manipulating Excel files.
- **win32com.client (win32)**: This library allows Python to interact with Outlook, enabling us to access emails and download attachments.
- **desktop_path**: Dynamically retrieves the user's desktop path using the USERPROFILE environment variable.
- **downloaded_folder_path**: This is where email attachments will be saved after they are downloaded.
- **os.makedirs()**: This function creates the folder if it doesn't already exist, ensuring that the script has a place to store downloaded attachments.

Summary

In this step, we imported the necessary libraries and defined the paths where email attachments will be saved. We also ensured that a folder exists to store downloaded attachments.

Step 2: Initializing Outlook and Accessing the Inbox

Next, we initialize the Outlook application and access the inbox to retrieve messages:

```python
# Initialize Outlook application
outlook = win32.Dispatch('outlook.application')
namespace = outlook.GetNamespace('MAPI')

# Access the inbox
inbox = namespace.GetDefaultFolder(6)  # 6 refers to the inbox folder
messages = inbox.Items
```

Explanation

- **win32.Dispatch('outlook.application')**: This line initializes a connection to Outlook, allowing us to interact with the application programmatically.
- **GetNamespace('MAPI')**: Accesses the MAPI (Messaging Application Programming Interface) namespace, which allows us to interact with folders, messages, and attachments in Outlook.
- **GetDefaultFolder(6)**: Retrieves the inbox folder (identified by the number 6 in MAPI) so we can access the emails stored there.
- **messages**: Stores all the messages in the inbox for further processing.

Summary

We connected to Outlook, accessed the inbox, and stored the messages for further processing. This allows us to interact with the email content and attachments directly from Python.

Step 3: Processing Emails and Downloading Attachments

In this step, we loop through the inbox messages, download the relevant attachments, and load them into pandas DataFrames:

```python
# Initialize a list to store DataFrames
dataframes = []

# Process each email in the inbox
for message in messages:
    if message.Subject.startswith("Monthly Report - ") and message.Attachments.Count > 0:
        # Download each attachment in the email
        for attachment in message.Attachments:
            file_name = attachment.FileName
            save_path = os.path.join(downloaded_folder_path, file_name)
            attachment.SaveAsFile(save_path)
            print(f"Downloaded: {file_name} to {downloaded_folder_path}")

            # Check if the file is an Excel file
            if file_name.endswith('.xlsx'):
                # Load the Excel file into a DataFrame
                df = pd.read_excel(save_path)
                dataframes.append(df)

```

Explanation

- **dataframes**: This list will store all the pandas DataFrames created from the downloaded Excel files.
- **message.Subject.startswith("Monthly Report - ")**: Checks if the email subject starts with "Monthly Report - ", which helps us filter out irrelevant emails.

- **message.Attachments.Count > 0**: Ensures the email has attachments before attempting to download anything.
- **attachment.SaveAsFile()**: Saves the attachment to the specified folder.
- **pd.read_excel()**: Loads the downloaded Excel file into a pandas DataFrame, which is then stored in the dataframes list for later aggregation.

Summary

We processed each email, downloaded attachments that match specific criteria, and loaded any Excel files into pandas DataFrames. This step is the heart of the script, as it automates the process of collecting relevant data from Outlook emails.

Step 4: Combining DataFrames and Saving the Aggregated Report

Finally, we combine all the DataFrames into a single report and save the result as a new Excel file:

```
# Concatenate all DataFrames into a single DataFrame
if dataframes:
    combined_df = pd.concat(dataframes, ignore_index=True)
    combined_df_path = os.path.join(downloaded_folder_path, 'Combined_Report.xlsx')
    combined_df.to_excel(combined_df_path, index=False)
    print(f"Combined DataFrame saved to: {combined_df_path}")
else:
    print("No relevant emails with attachments found.")
```

Explanation

- **pd.concat()**: Concatenates all the DataFrames in the dataframes list into a single DataFrame.
- **ignore_index=True**: Ensures that the index is reset in the final DataFrame, making it clean and sequential.
- **combined_df.to_excel()**: Saves the concatenated DataFrame to a new Excel file, Combined_Report.xlsx.
- **index=Fale**: The index column is omitted from the downloaded excel file.

Summary

We combined all the DataFrames collected from email attachments into one consolidated DataFrame and saved it as an Excel file. This step completes the automation of collecting, processing, and aggregating reports from Outlook.

Final Output

▶ After running this code, reports attached to specific emails in your Outlook inbox will be automatically downloaded, processed, and combined into a single Excel file. Here's a breakdown of what the code accomplishes:

1. **Folder Creation for Downloaded Files**: The script creates an **"Email Downloaded Folder"** within the **"Automate_Everything_With_Python"** directory on your desktop. This is where all downloaded attachments will be saved, ensuring organized storage of files related to this automated process.

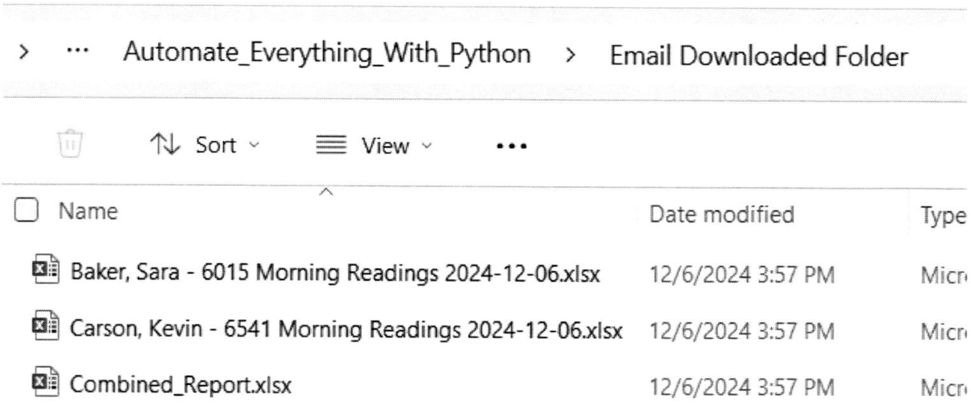

2. **Connecting to Outlook and Accessing Inbox**: The script initializes a connection to your Outlook inbox, scanning all emails in the inbox to identify those that meet specific criteria.
3. **Filtering Relevant Emails and Downloading Attachments**: The script looks for emails with a subject line starting with **"Monthly Report - "** and that contain attachments. For each qualifying email:

    a. It downloads all attachments to the **"Email Downloaded Folder"** on your desktop.
    b. Each downloaded file is confirmed with a print message, allowing you to track which files were saved.

4. **Processing and Aggregating Excel Files**: After downloading, the script checks if each attachment is an Excel file (.xlsx):

    a. If it's an Excel file, it loads the data into a pandas DataFrame.

b. All DataFrames from the downloaded files are then combined into a single DataFrame.

5. **Saving the Combined Report**: Once all relevant files are processed, the script saves the combined data into a single Excel file named **"Combined_Report.xlsx"** within the **"Email Downloaded Folder"**. A message confirms the successful creation of the combined report file.
6. **Notification if No Relevant Emails Are Found**: If no emails with the specified subject line and attachments are found, the script prints a message informing you that no relevant emails were available for processing.

Benefits of This Automated Report Collection and Aggregation

By executing this code, you:

- **Streamline Report Collection**: Automate the tedious process of downloading attachments from multiple emails, making report collection faster and more efficient.
- **Ensure Organized Data Storage**: The downloaded files and the combined report are saved in a designated folder, maintaining a clean and organized workspace.
- **Create an Aggregate Report Effortlessly**: The code consolidates multiple Excel files into a single report, saving you time and reducing the risk of errors associated with manual data aggregation.

This script is especially useful for businesses or teams that receive recurring reports by email. By automating the collection and aggregation process, you'll have an up-to-date, combined report readily available, supporting efficient data analysis and reporting workflows.

Conclusion

Congratulations! You've successfully automated the process of collecting and aggregating reports from Outlook. Here's a recap of what you've learned:

- Interacting with Outlook using the win32com library to access emails and download attachments.
- Downloading and processing email attachments by filtering emails based on subject lines and attachment types.
- Using pandas to load and manipulate Excel files, allowing you to aggregate multiple reports into a single comprehensive file.

Your script is now a powerful tool that can:

- Automatically identify and download relevant reports from Outlook emails.
- Load, process, and aggregate data from multiple Excel files into a single report.
- Save time by eliminating the need for manual report collection and combination.

By mastering these techniques, you're well-equipped to streamline report collection and aggregation in your organization, saving time and improving data accuracy.

Chapter 10: Creating Consolidated Text Backup of Python Scripts

Introduction to Creating Consolidated Text Backup of Python Scripts

Welcome back! In this chapter, we'll focus on automating the process of creating a consolidated backup of all Python scripts in a directory. By the end of this chapter, you will know how to automatically gather all Python files, concatenate their contents, and save them into a single text file with a date stamp. This is especially useful for managing backups, reviewing all scripts at once, or sharing code as a single file.

Code Overview: Automating the Consolidated Backup

Before we dive into the details, here's a high-level overview of the script we'll build:

- Setting up the environment: Importing libraries and defining file paths.
- Creating a backup folder: Ensuring a dedicated space for backups exists.
- Gathering Python files: Filtering the list to include only .py files.
- Consolidating Python files: Concatenating the contents of all .py files into a single text file.
- Final output: Saving the backup text file with a date-stamped filename.

Let's now break this process into detailed steps.

```
# ============================================================================
# Chapter 10 - Creating Consilidated Text Backup of Python Script.py
# ============================================================================

import pandas as pd
import os
import datetime as datetime

desktop_path = os.path.join(os.environ['USERPROFILE'], 'Desktop')
automate_everything_path = os.path.join(desktop_path, 'Automate Everything With Python')
python_code_path = os.path.join(desktop_path, 'Automate Everything With Python', 'Python Code')
backup_folder_path = os.path.join(desktop_path, 'Automate Everything With Python','Python Backup Files')

# Create the backup folder if it doesn't exist
if not os.path.exists(backup_folder_path):
    os.makedirs(backup_folder_path)

# Create a subfolder with the current date
current_date = datetime.datetime.now().strftime('%Y-%m-%d')
date_folder_path = os.path.join(backup_folder_path, current_date)



##### Gather all files #
all_py_files = os.listdir(python_code_path)

# remove unwanted files
all_py_files = [x for x in all_py_files if '.py' in x]
# allfiles = [x for x in allfiles if 'Completed' not in x]

##### iterate on all files and append to dataframe #
for f in all_py_files:
    with open(python_code_path + '\\' + f,'r') as py_file:
        py_content = py_file.read()
        with open(backup_folder_path + '\\' + 'Python Log ' + current_date + '.txt','a') as txt_file:
            txt_file.write(py_content)

print('Catologing Python script is finished!!!')

```

Step 1: Importing Libraries and Setting Up File Paths

First, we need to import the necessary libraries and define file paths for the script. The code below sets up the environment and ensures we have paths for both the Python code and backup folders.

```
import pandas as pd
import os
import datetime as datetime
```

Explanation

- **pandas**: Imported for future data manipulation or handling needs.

- **os**: Helps manage file and directory operations.
- **datetime**: Used for generating the current date for timestamping the backup files.

Next, we define the paths for the desktop, project folder, Python code directory, and backup folder:

```
desktop_path = os.path.join(os.environ['USERPROFILE'], 'Desktop')
automate_everything_path = os.path.join(desktop_path, 'Automate_Everything_With_Python')
python_code_path = os.path.join(desktop_path, 'Automate_Everything_With_Python', 'Python Code')
backup_folder_path = os.path.join(desktop_path, 'Automate_Everything_With_Python','Python Backup Files')
```

Explanation

- **desktop_path**: Gets the path to the desktop using environment variables.
- **automate_everything_path**: Defines the project folder path on the desktop.
- **python_code_path**: Points to the folder where the Python scripts are stored.
- **backup_folder_path**: Specifies the folder where backups will be saved.

Summary

- We imported essential libraries (pandas, os, and datetime).
- Defined paths for the desktop, project folder, script folder, and backup folder.
- This setup ensures flexibility and that the script works across different machines.

Step 2: Creating the Backup Folder

In this step, we create the backup folder and a subfolder named with the current date. This is critical for organizing backups and avoiding file overwrites.

```
# Create the backup folder if it doesn't exist
if not os.path.exists(backup_folder_path):
    os.makedirs(backup_folder_path)
```

Explanation

- **os.path.exists()**: Checks if the backup folder already exists.
- **os.makedirs()**: Creates the folder if it does not already exist, ensuring a dedicated space for backups.

Now, we create a subfolder with the current date, formatted as YYYY-MM-DD, to store the backup:

```
# Create a subfolder with the current date
current_date = datetime.datetime.now().strftime('%Y-%m-%d')
date_folder_path = os.path.join(backup_folder_path, current_date)
```

Explanation

- **datetime.datetime.now()**: Gets the current date and time.
- **strftime('%Y-%m-%d')**: Formats the date to ensure consistency (e.g., 2023-10-18).
- **date_folder_path**: Defines a path for the subfolder using the current date for better organization.

Summary

- Checked if the backup folder exists and created it if necessary.
- Organized the backups by date, ensuring that each backup is stored in a new, separate subfolder.

Step 3: Gathering and Filtering Python Files

We now gather all the files from the Python Code folder and filter out non-Python files, ensuring we only back up .py files.

```
##### Gather all files #
all_py_files = os.listdir(python_code_path)
```

Explanation

- **os.listdir()**: Lists all files in the python_code_path directory.

Next, we filter out non-Python files:

```
# remove unwanted files
all_py_files = [x for x in all_py_files if '.py' in x]
```

- **List comprehension**: Iterates over all the files, selecting only those that end with .py.

Summary

- Gathered all files from the Python Code directory.
- Filtered the list to include only Python files (.py).

Step 4: Consolidating Python Files into a Single Text File

This step involves reading the content of each Python file and writing it into a single text file, creating a consolidated backup.

```
##### iterate on all files and append to dataframe #
for f in all_py_files:
    with open(python_code_path + '\\' + f,'r') as py_file:
        py_content = py_file.read()
        with open(backup_folder_path + '\\' + 'Python Log ' + current_date + '.txt','a') as txt_file:
            txt_file.write(py_content)
```

Explanation

- **Iteration**:
  - Loops through each Python file in all_py_files.
- **File Reading**:
  - **with open(python_code_path + '\\' + f, 'r')**: Opens each Python file for reading.
  - **py_content = py_file.read()**: Reads the content of the file and stores it in the py_content variable.
- **Writing to Backup File**:
  - **with open(backup_folder_path + '\\' + 'Python Log ' + current_date + '.txt', 'a')**: Opens the backup text file in append mode. If the file doesn't exist, it will be created.
  - **txt_file.write(py_content)**: Writes the content of the Python file to the backup file, appending each script's contents.

Summary

- Iterated through all Python files in the directory.
- Concatenated each Python file's content into a single text file.
- Ensured that the consolidated backup is saved with the current date in the filename.

Step 5: Completing the Process

Finally, we add a message to indicate that the backup process is complete.

```

print('Catologing Python script is finished!!!')

```

Explanation

This provides feedback that the process has successfully completed, and the consolidated backup has been created.

Summary

- Printed a message confirming that the backup process is complete.

After running this code, all Python scripts in the "Python Code" folder will be cataloged into a single log file. Here's a step-by-step breakdown of what the script accomplishes:

1. Backup Folder Setup: The script checks if the "Python Backup Files" folder exists within the "Automate_Everything_With_Python" directory on your desktop. If it doesn't, the script creates the folder, ensuring there's a designated space for backup and log files.
2. Date-Stamped Subfolder Creation: A new subfolder named with the current date (formatted as YYYY-MM-DD) is created within the "Python Backup Files" folder. This date-stamped folder provides an organized location for storing logs based on when they were created.

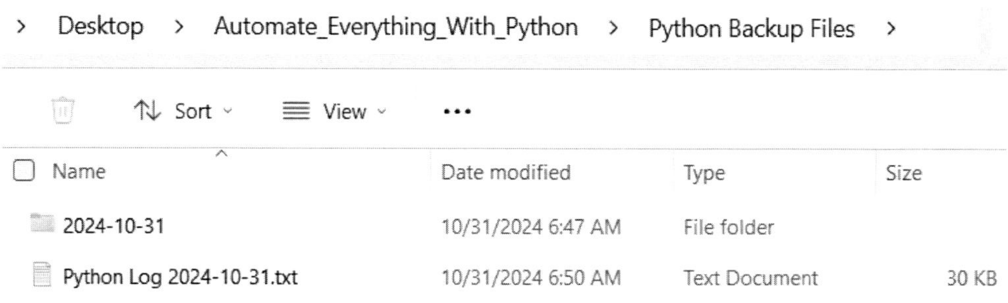

3. Identifying Python Files: The script scans the "Python Code" folder and gathers a list of all Python files (files ending with .py). This list includes only the relevant files, filtering out any non-Python files.

4. Cataloging Python Code into a Log File: For each Python file in the "Python Code" folder:

    a. The script opens the file, reads its contents, and writes them into a log file.
    b. This log file, named "Python Log YYYY-MM-DD.txt", is stored in the "Python Backup Files" folder with the current date in the filename. Each script's contents are appended to this log file, creating a consolidated view of all Python scripts in one document.

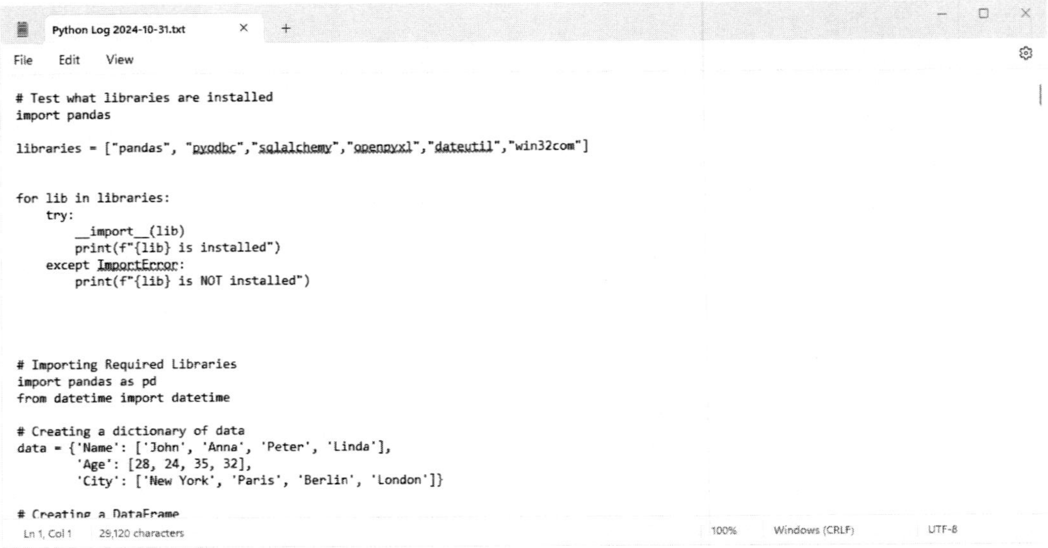


```
# Test what libraries are installed
import pandas

libraries = ["pandas", "pyodbc","sqlalchemy","openpyxl","dateutil","win32com"]


for lib in libraries:
    try:
        __import__(lib)
        print(f"{lib} is installed")
    except ImportError:
        print(f"{lib} is NOT installed")



# Importing Required Libraries
import pandas as pd
from datetime import datetime

# Creating a dictionary of data
data = {'Name': ['John', 'Anna', 'Peter', 'Linda'],
        'Age': [28, 24, 35, 32],
        'City': ['New York', 'Paris', 'Berlin', 'London']}

# Creating a DataFrame
```

Benefits of This Cataloging Process

By executing this code, you:

- **Create a Comprehensive Record of Python Code**: The log file serves as a consolidated reference for all Python scripts, which can be useful for code review, version tracking, or auditing.
- **Ensure Organized, Date-Stamped Backups**: The date-stamped naming convention allows for easy retrieval of backups, making it clear when each cataloging process was performed.
- **Enable Quick Code Access in One File**: Having all scripts in one log file makes it easy to search for code snippets, functions, or references without opening each file individually.

This script offers a simple yet effective way to archive and track changes to Python code, supporting version control and code maintenance efforts.

Conclusion

In this chapter, you've learned how to automate the process of creating a consolidated backup of Python scripts. Here's what we covered:

- **Set up the environment**: Imported libraries and defined file paths.
- **Created the backup folder**: Ensured backups are stored in an organized, date-stamped subfolder.
- **Gathered and filtered Python files**: Selected only .py files for backup.
- **Consolidated files**: Read each Python file and appended its content to a consolidated text file.
- **Completed the process**: Added a confirmation message when the backup is finished.

This script simplifies the process of managing and reviewing multiple Python scripts by consolidating them into a single file. As a next step, you can enhance this script by adding features such as error handling, logging, or even compressing the backup file to save space.

By mastering these techniques, you've gained valuable skills to streamline your backup process and ensure your Python projects are well-organized and easily accessible. Keep practicing, and soon you'll be an expert in automating your development tasks!

Chapter 11:
Automating Excel Macros

Introduction to Excel Macro Automation

Welcome back, automation virtuosos! We're about to embark on an exciting journey that will revolutionize how you interact with Excel macros. In this chapter, we'll craft a powerful Python script that automates the execution of Excel macros. This is a game-changing technique for business analysts and data professionals who work with complex Excel models containing VBA macros.

By the end of this chapter, you'll be able to open an Excel workbook, run a specific macro, and then save and close the workbook, all without manually opening Excel. This skill is particularly valuable for automating repetitive tasks, integrating Excel-based processes into larger Python workflows, and running time-consuming Excel operations in the background. Get ready to become the master of Excel automation in your data processing and analysis workflows!

Understanding the Excel File and VBA Macro

Before we dive into our Python script, let's take a moment to understand the Excel file and the macro we'll be working with:

1. **Open the Excel File:**
    a. Navigate to the folder: Desktop\Automate_Everything_With_Python\Excel Models
    b. Open the file named "PCP Compensation Model.xlsm"
2. **Locate the Macro:**
    a. Press Alt + F11 to open the Visual Basic Editor
    b. In the Project Explorer, look for "Modules"
    c. Double-click on the module to view its contents
    d. You should see the macro named "GenerateStatements"
3. **Macro Name:**
    a. The macro we're automating is named "GenerateStatements"
    b. This is the name we use in our Python script: excel.Application.Run("'PCP Compensation Model.xlsm'!GenerateStatements")

Explanation of the VBA Macro

The GenerateStatements macro automates the process of generating PDF reports for multiple individuals based on data in the Excel workbook. Here's a breakdown of its key functions:

1. **Initialization:**
    a. Turns off screen updating for faster execution
    b. Sets up variables for looping through data and creating file paths

2. **File Path Setup:**

    a. Creates a folder on the desktop: \Automate_Everything_With_Python\Automated Reports\yyyy-mm Reports
    b. The folder name includes the current year and month

3. **PDF Generation Loop:**

    a. Starts with the first person in the list (set in cell B2 of the "vba" sheet)
    b. Exports the "CompModel" sheet as a PDF for each person
    c. The PDF filename includes the person's name (from cell H2 of the "CompModel" sheet)
    d. Loops through all entries until it reaches the maximum value (set in cell B1 of the "vba" sheet)

4. **Key Features:**

    a. Uses ExportAsFixedFormat to create PDFs
    b. Dynamically names files based on worksheet data
    c. Creates a new folder for each month's reports
    d. Automates the process for multiple individuals without manual intervention

Understanding this VBA code helps us appreciate what our Python script is automating. It's taking a process that would typically require opening Excel, running the macro manually, and waiting for it to complete, and turning it into a single, quick operation that can be run from Python without even opening Excel visually.

Code Overview: Automating Excel Macros with Python

The script we'll build covers the following steps:

- **Importing libraries**: Interacting with Excel via Python.
- **Setting up file paths**: Pointing to the Excel workbook.
- **Launching Excel in the background**: Running Excel without visual interaction.
- **Running the macro**: Executing the desired Excel macro.
- **Saving and closing the workbook**: Finalizing the operation by saving any changes.
- **Quitting Excel**: Freeing system resources.

Now, let's dive into the details step by step.

```python
import os
import win32com.client as win32

# Define the path to the workbook
desktop_path = os.path.join(os.environ['USERPROFILE'], 'Desktop')
excel_models_path = os.path.join(desktop_path, 'Automate Everything With Python', 'Excel Models')
workbook_path = os.path.join(excel_models_path, 'PCP Compensation Model.xlsm')

# Initialize Excel application
excel = win32.Dispatch('Excel.Application')
excel.Visible = False  # Run Excel in the background

# Open the workbook
workbook = excel.Workbooks.Open(workbook_path)

# Run the macro
excel.Application.Run("'PCP Compensation Model.xlsm'!GenerateStatements")

# Save and close the workbook
workbook.Save()
workbook.Close()

# Quit Excel
excel.Quit()

print(f"Macro 'GenerateStatements' executed successfully in {workbook_path}.")
```

Step 1: Importing Libraries and Setting Up File Paths

First, let's import the required libraries and set up file paths for our Excel workbook.

```python
import os
import win32com.client as win32
```

Explanation

- **os**: Used for file and directory operations.
- **win32com.client**: This allows Python to interact with COM objects like Excel.

Next, define the paths to the desktop and the Excel workbook you want to automate:

```python
# Define the path to the workbook
desktop_path = os.path.join(os.environ['USERPROFILE'], 'Desktop')
excel_models_path = os.path.join(desktop_path, 'Automate_Everything_With_Python', 'Excel Models')
workbook_path = os.path.join(excel_models_path, 'PCP Compensation Model.xlsm')
```

Explanation

- **desktop_path**: Fetches the path to the user's desktop.
- **excel_models_path**: Specifies the folder containing the Excel models.
- **workbook_path**: Points directly to the Excel workbook (PCP Compensation Model.xlsm) we want to automate.

Summary

- Imported the necessary libraries (os and win32com.client).
- Defined paths to the desktop, project folder, and the specific Excel workbook.

Step 2: Initializing the Excel Application

Next, we initialize an instance of Excel, but keep it invisible to avoid opening the application on screen.

```
# Initialize Excel application
excel = win32.Dispatch('Excel.Application')
excel.Visible = False  # Run Excel in the background
```

Explanation

- **win32.Dispatch('Excel.Application')**: Launches an instance of Excel.
- **excel.Visible = False**: Runs Excel in the background, preventing it from popping up.

Summary

- Launched an instance of Excel in the background to enable automation without user interaction.

Step 3: Opening the Workbook

Once Excel is initialized, the next step is to open the specific workbook that contains the macro.

```
# Open the workbook
workbook = excel.Workbooks.Open(workbook_path)
```

Explanation

- **Workbooks.Open()**: Opens the Excel workbook located at workbook_path and returns a reference to it.

Summary

- Opened the Excel workbook programmatically from the specified path.

Step 4: Running the Macro

Now that the workbook is open, we can run the Excel macro.

```
# Run the macro
excel.Application.Run("'PCP Compensation Model.xlsm'!GenerateStatements")
```

Explanation

- **Application.Run()**: Executes a specified macro from the workbook.
- **Syntax**: The argument 'PCP Compensation Model.xlsm!GenerateStatements' tells Excel to run the GenerateStatements macro inside the specified workbook.

Summary

- Executed the GenerateStatements macro from the PCP Compensation Model.xlsm workbook.

Step 5: Saving and Closing the Workbook

After running the macro, it's important to save any changes made by the macro and then close the workbook.

```
# Save and close the workbook
workbook.Save()
workbook.Close()
```

Explanation

- **workbook.Save()**: Saves the changes made to the workbook.
- **workbook.Close()**: Closes the workbook after saving.

Summary

- Saved any changes to the workbook and closed it properly.

Step 6: Quitting Excel and Printing Confirmation

Finally, we quit the Excel application and print a confirmation message indicating that the macro was successfully executed.

```
# Quit Excel
excel.Quit()

print(f"Macro 'GenerateStatements' executed successfully in {workbook_path}.")
```

Explanation

- excel.Quit(): Exits the Excel application and frees up system resources.
- **Print statement**: Outputs a confirmation message to let you know the macro ran successfully.

Summary

- Quit the Excel application and printed a confirmation message to confirm successful execution of the macro.

Step 7: Generating PDF Files and Organizing Them by Date

Once the macro is successfully run by the script, it automates a key process—generating PDF reports for multiple providers. The GenerateStatements macro, which was executed in **Step 4**, handles this by exporting reports into a folder that is automatically created by the macro. Here's how it works:

When you run the script, the following actions are taken by the macro:

1. **Creates a folder for the reports**: The macro dynamically creates a folder named with the **current year and month** inside the Automated Reports folder on your desktop. For example, if the script is run in October 2024, the folder will be named:

   C:\Users\YourUsername\Desktop\Automate_Everything_With_Python\Automated Reports\2024-10 Reports

2. **Generates PDFs for each provider**: The macro loops through a list of providers in the Excel workbook and generates a PDF for each provider. These PDF reports are exported to the newly created folder.
3. **Names the PDF files**: Each PDF file is named dynamically based on the provider's name. For example, if the provider's name is "Zachary Anderson" the corresponding PDF will be named something like:

   Anderson, Zachary - 6535 - PCP Compensation Model.pdf

What to Expect After Running the Script

After running the Python script, you will find that:

- A new folder with the current year and month has been created inside the Automated Reports folder.
- Inside this folder, there will be PDF files for each provider, named according to the provider's details.

This process eliminates the need to manually export reports, ensuring that everything is organized in a date-stamped folder for easy reference.

Example:

If you run the script in October 2024, and providers named Zachary Anderson and Stephanie Collins, the following files and folders will be created:

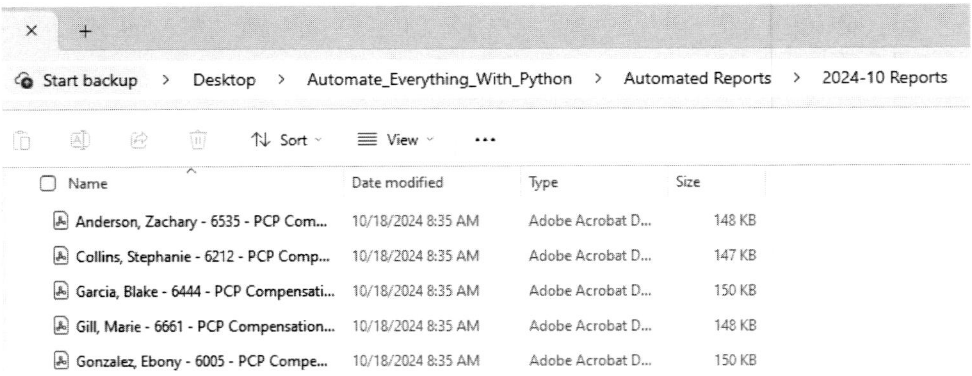

This structure allows you to efficiently store and retrieve reports, with the added benefit of automatic date-based organization. Each time you run the script in a new month, a new folder will be created, keeping your reports well-organized without any manual effort.

Summary

- After running the code, a folder with the current year and month is created in the Automated Reports directory.
- Inside this folder, PDF files are generated for each provider, named appropriately.

- This structure ensures organized storage of reports, making it easy to track reports by month and year.

Final Output

▶ After running the Python code, the following actions will take place:

1. **Workbook Path Setup**: The code defines the path to the "PCP Compensation Model.xlsm" workbook located in the "Excel Models" folder within the "Automate_Everything_With_Python" directory on the desktop.

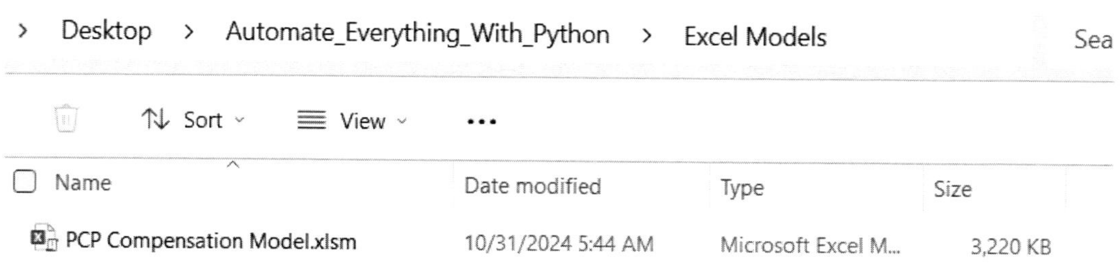

2. **Excel Initialization**: The script launches Microsoft Excel in the background without displaying the Excel application window. This helps the code run efficiently and without user interaction.

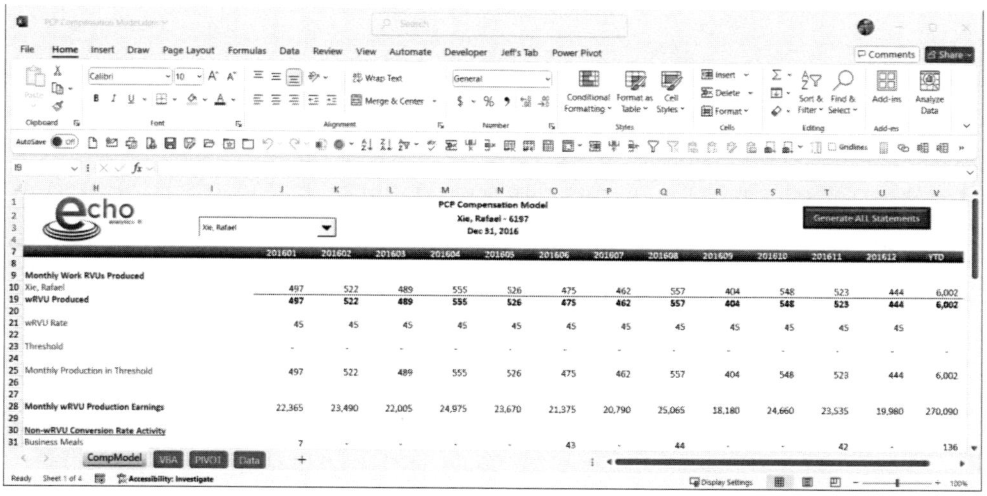

3. **Opening the Workbook and Running the Macro**: The Python script opens the specified Excel workbook and runs the "GenerateStatements" macro within it. This macro automates the process of generating PDF reports for individuals listed in the "vba" sheet of the workbook.
4. **Generating Folder Paths**:
   a. The macro creates a folder structure on the desktop within the "Automate_Everything_With_Python\Automated Reports" directory.

b. A subfolder is created with the current period in YYYY-MM format (e.g., "2024-10 Reports") to store the PDF reports for the current month.

5. **Exporting the First PDF Report**:
   a. The macro exports the "CompModel" worksheet as a PDF file for the first person in the list.
   b. The PDF file is named based on the person's name or ID located in cell H2 of the "CompModel" worksheet, followed by " - PCP Compensation Model.pdf".
   c. The PDF is saved in the newly created period folder, maintaining a clear and consistent naming convention.

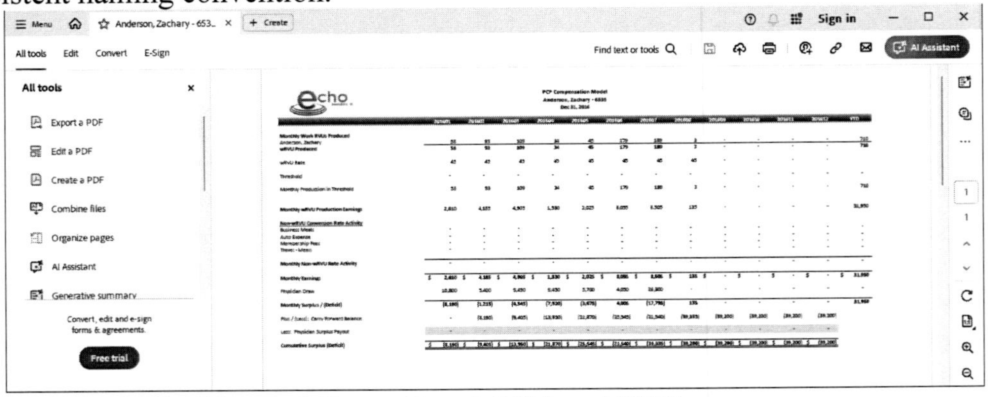

6. **Looping Through the List and Exporting Additional PDFs**:
   a. The macro increments through each person in the list on the "vba" sheet.
   b. For each person, the "CompModel" worksheet is exported as a PDF with the same naming structure, saving it in the period folder.
   c. This loop continues until PDF files are created for all individuals listed in the "vba" sheet.

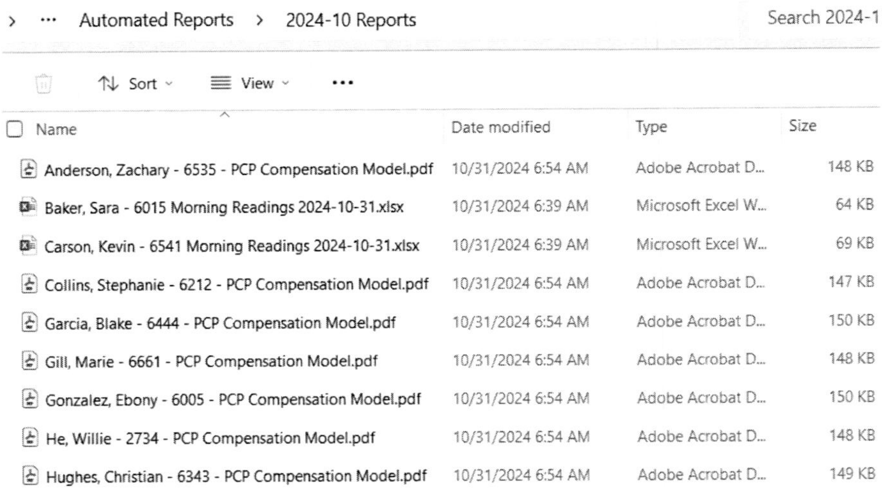

Conclusion

In this chapter, we automated the execution of an Excel macro using Python. Here's a recap of what we covered:

- **Imported libraries**: Set up Python to interact with Excel using the win32com.client library.
- **Defined file paths**: Pointed to the Excel workbook you want to automate.
- **Initialized Excel**: Ran Excel in the background without any visual interaction.
- **Ran the macro**: Automated the process of running the GenerateStatements macro.
- **Saved and closed the workbook**: Ensured any changes were saved, and the workbook was closed properly.
- **Quit Excel**: Freed up system resources and confirmed successful execution.

This script is a powerful tool to automate repetitive tasks in Excel. You can further enhance this script by:

- Adding error handling to manage cases where the workbook or macro does not exist.
- Running macros in multiple workbooks in a loop.
- Logging the execution to track the history of macro runs.

By mastering these techniques, you can integrate Excel into larger Python workflows, saving time and improving efficiency.

Chapter 12: Automating Python with Task Scheduler

Introduction to Automating Python with Task Scheduler

Welcome back, automation enthusiasts! In this chapter, we will explore how to automate Python script execution using Windows Task Scheduler. This technique allows you to schedule your Python scripts to run at predetermined times or intervals without any manual intervention. This is particularly useful for automating repetitive tasks like data processing, reporting, or system maintenance.

By the end of this chapter, you'll be able to set up Task Scheduler to run your Python scripts automatically. This powerful tool will significantly enhance your workflow, helping you become more efficient and productive.

Step 1: Writing the Python Code

Code Overview: Locating Your Python Executable: A Critical Step for Automation

Before diving into automation tasks, it's essential to understand where your Python executable is located on your system. Knowing this location is particularly useful when configuring automation tools, setting up environment variables, or scheduling tasks. The following Python code will help you locate the Python executable that your scripts will use.

```
import sys

# Get the path to the Python executable
python_executable = sys.executable

print(f"The Python executable is located at: {python_executable}")

# Optional: print the version of Python
print(f"Python version: {sys.version}")
```

Explanation

- **sys**: The sys module provides tools for interacting with the Python interpreter and the system environment.
- **sys.executable**: This attribute retrieves the absolute path to the Python executable currently running the script.
- **print()**: Displays the path to the Python executable for you to copy and use.
- **sys. version**: Prints the Python version to confirm which Python interpreter is being used.

Final Output

▶ When you run the script, the console will display essential information needed to configure your Python scripts for automation in Windows Task Scheduler. Specifically, it will show the file path to the Python executable, which is required to set up the Task Scheduler. For example, you might see: The Python executable is located at: C:\Users\YourName\AppData\Local\Programs\Python\Python39\python.exe. Along with this, the script will output the version of Python associated with the executable to confirm that the correct version is being used. An example output could be: Python version: 3.9.10 (default, Mar 16 2023, 15:39:10) [MSC v.1928 64 bit (AMD64)]. A screenshot of this output is provided below to ensure you know exactly what to expect when running the code. Make sure to copy the Python executable path from the output for use in Task Scheduler.

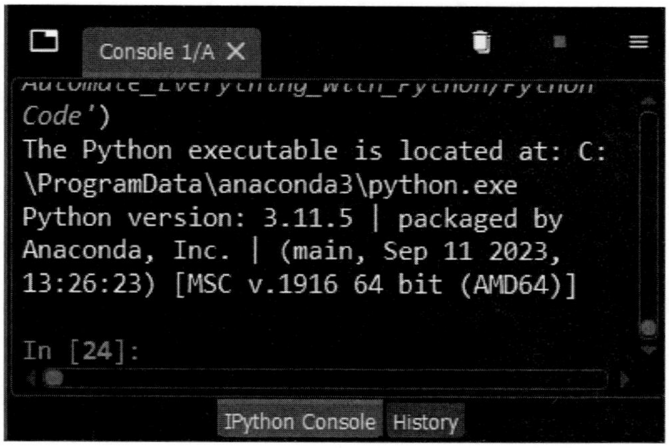

Step 2: Setting Up Task Scheduler

Now, we'll set up Task Scheduler to automatically run the Python script.

1. **Opening Task Scheduler**
   a. **Open Task Scheduler**:
      - Press **Win + R**, type taskschd.msc, and press **Enter**. This will open the Task Scheduler interface.

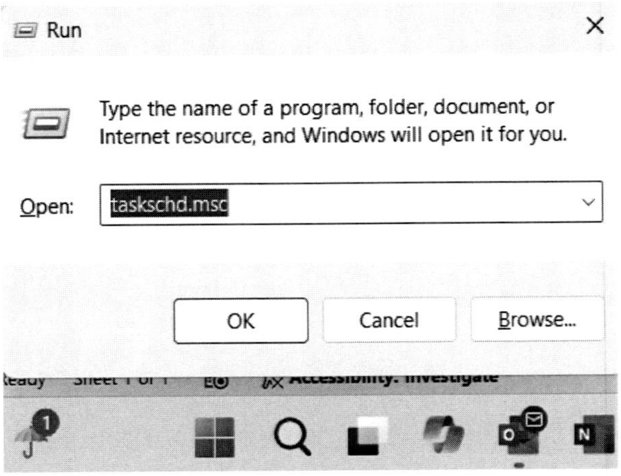

2. **Creating a New Task**
   a. **Create a Basic Task**:
      - In the right-hand pane under **Actions**, click on **Create Basic Task...**.

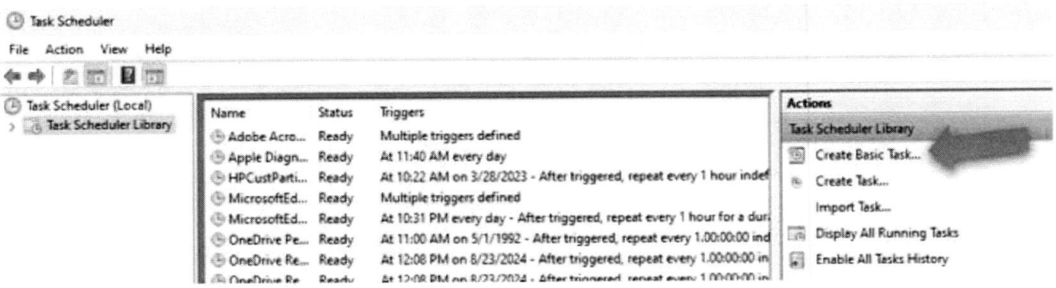

      - Enter a name for your task, such as Run Python Script.
      - Optionally, add a description to clarify the purpose of the task.
      - Click **Next**.

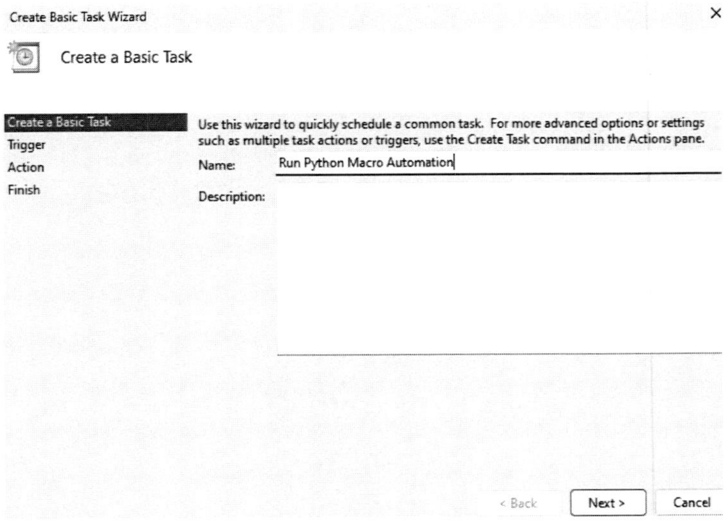

3. **Configuring the Task Trigger**
   a. **Set the Trigger**:
      - Choose **One time** as the trigger type for this example (you can schedule it for Daily, Weekly or Monthly intervals later if desired).
      - Click **Next** and set the date and time for when you want the task to run.
      - Click **Next.**

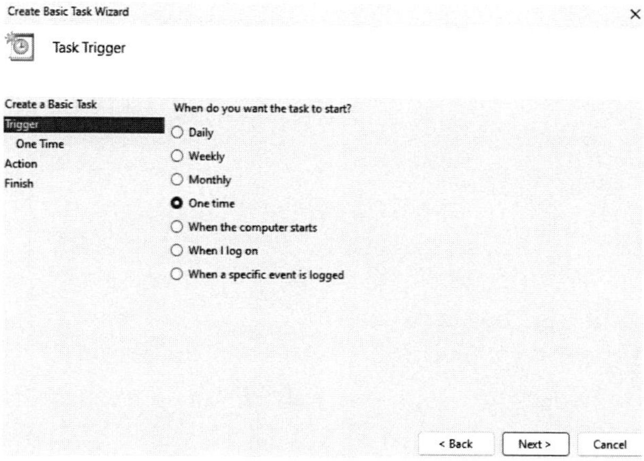

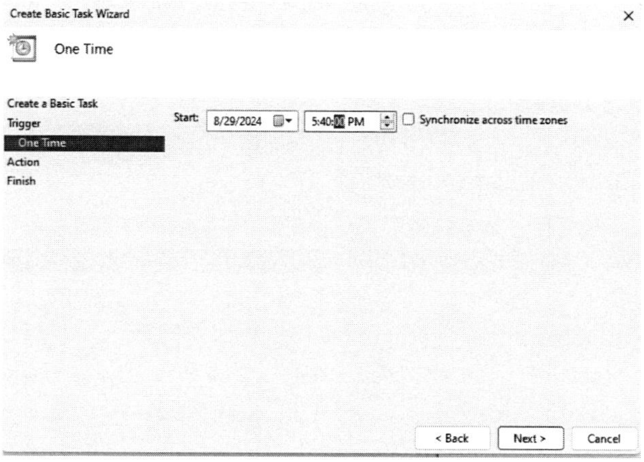

4. **Configuring the Task Action**
   a. **Set the Action**:
      - Select **Start a program** and click **Next**.

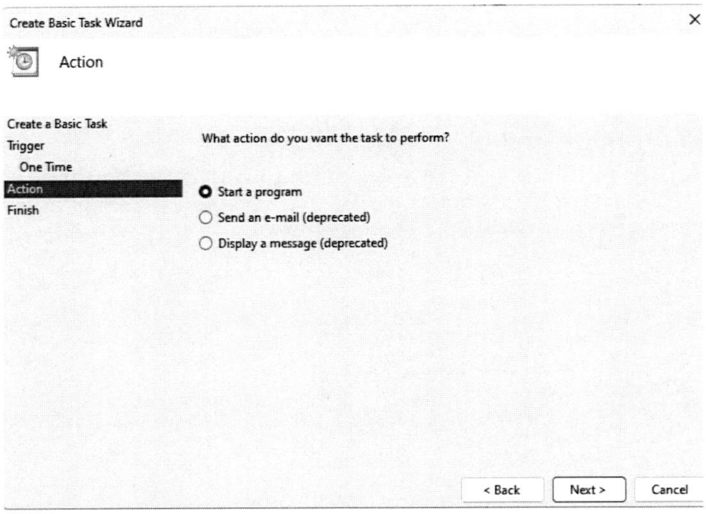

5. **Configuring the Start a Program**
   a. **Specify the Python Executable**:
      - In the **Program/script** field, paste the **Python executable path** that you retrieved earlier using the code created in Spyder (enclose in double quotes), for example:

```
C:\ProgramData\anaconda3\python.exe
```

   b. **Add the Python Script Path**:
      - In the **Add arguments (optional)** field, enter the full path to your Python script (in double quotes), for example:

```
"C:\Users\YourUsername\Desktop\Automate_Everything_With_Python\Python Code\Chapter 11 - Automating Excel Macros.py"
```

   c. **Click Next** to continue.

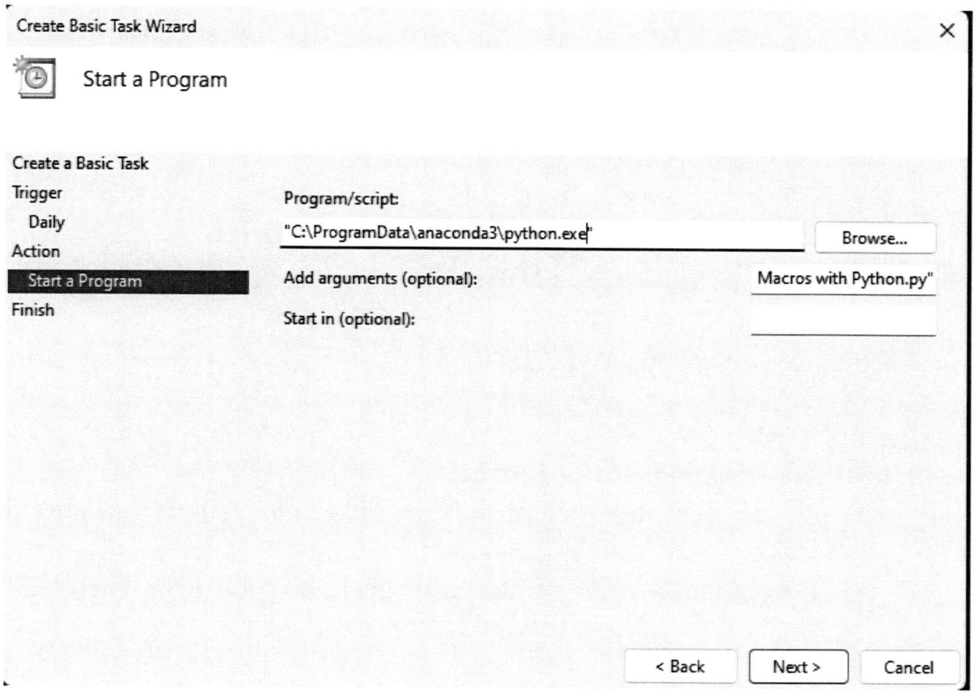

6. **Reviewing and Finishing the Task**
   a. **Review the Task**:
      - Review the summary of the task to make sure everything is set up correctly.
      - Click **Finish** to create the task.

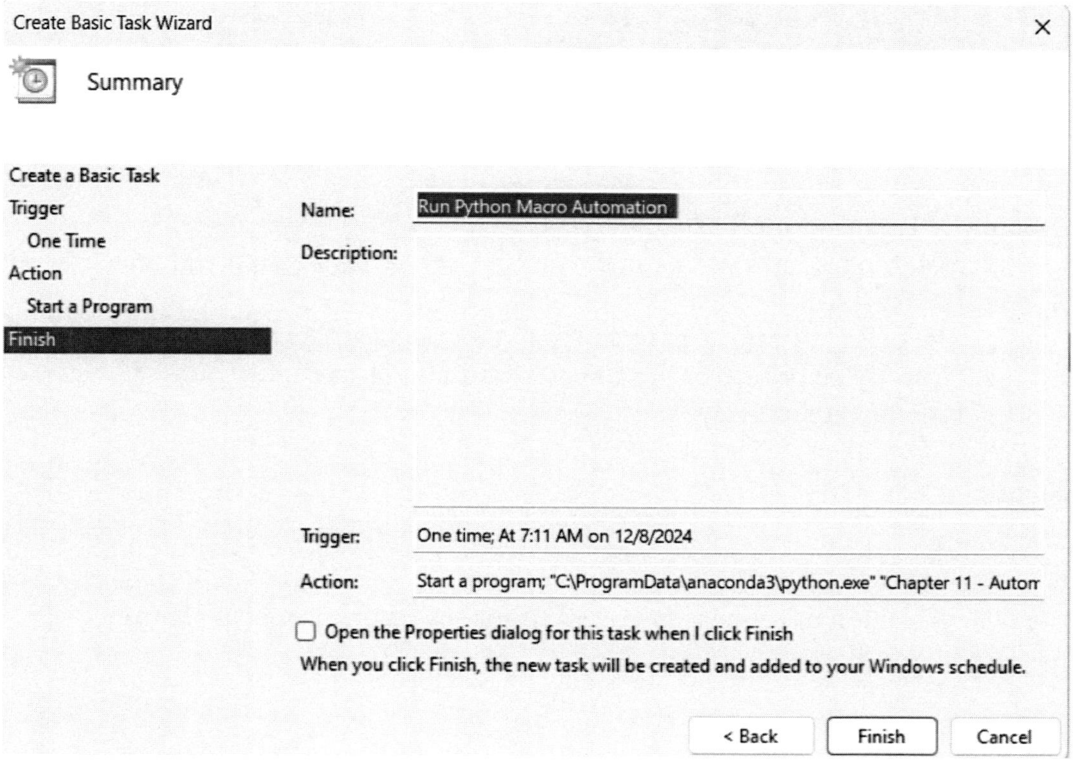

Summary

Configured Task Scheduler to automatically run the Python script at a specific time, ensuring that the script is executed with the correct Python interpreter.

Step 3: Monitoring Task Execution

Once the task is set up, Task Scheduler will run the Python script at the designated time.

1. **Verifying the Task**
   a. **Check the Task Scheduler Library**:
      - In Task Scheduler, click on **Task Scheduler Library** to view the list of tasks.
      - Locate the task you created and check the **Next Run Time** column to verify when it is scheduled to run.

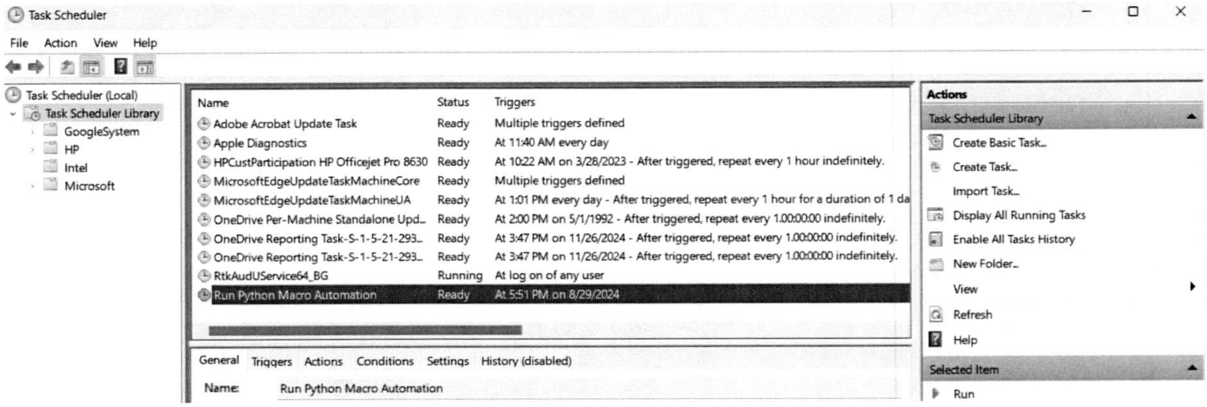

2. **Manually Running the Task**:
   a. You can also manually run the task by right-clicking on it and selecting **Run**.
3. **Check the Output**:
   a. The output of the script (e.g., the Python executable path and version) will typically be printed in the command prompt or console window if the task is running interactively.

Summary

Verified that the task is scheduled correctly and can be manually executed from Task Scheduler. Monitored the execution of the task and checked for any errors using the **History** tab.

Final Output

When Task Scheduler executed the Python script, the following actions took place:

1. **Workbook and Macro Setup**: The script located the "PCP Compensation Model.xlsm" workbook and ran the "GenerateStatements" macro.

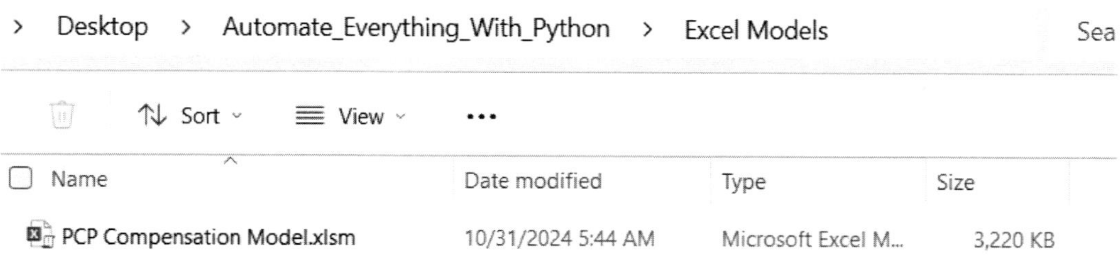

2. **PDF Generation**: The macro created personalized PDF reports for each individual listed in the workbook, starting with the first person and looping through the list.
3. **Folder Creation**: A date-stamped folder (e.g., "2024-10 Reports") was automatically generated in the "Automated Reports" directory to store the PDFs.
4. **File Naming and Organization**: Each PDF was named using a structured format (e.g., "[Name/ID] - PCP Compensation Model.pdf") and saved in the appropriate folder.

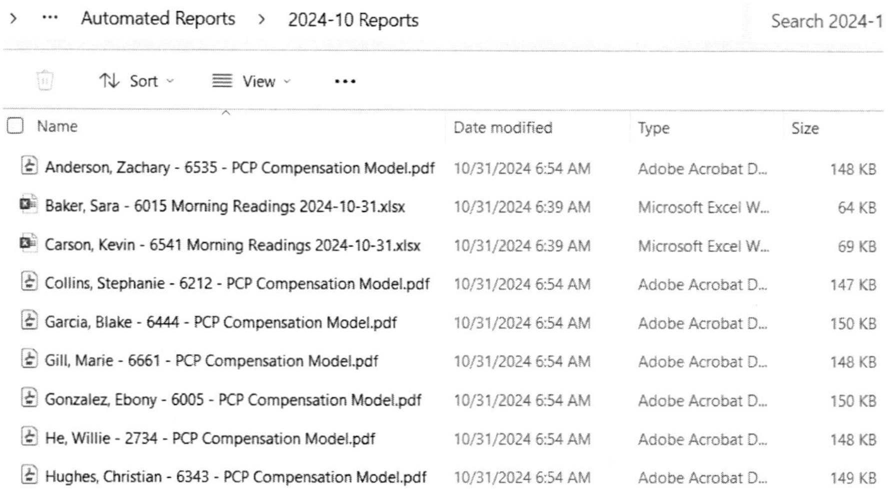

This automated process ensured all reports were generated, named, and stored systematically, without requiring manual intervention.

Benefits of Task Scheduler Automation

By using Task Scheduler to execute this Python script:

- **Hands-Free Execution**: The entire process runs automatically at scheduled times, without needing manual intervention.
- **Consistent and Timely Report Generation**: Task Scheduler ensures that reports are generated and saved on a regular basis, following a structured naming and storage convention.
- **Efficient Report Organization**: Each report is stored in a date-stamped folder, making it easy to track reports by month and maintain an organized archive.

This setup allows you to automate recurring reporting tasks reliably, freeing up time and ensuring reports are consistently generated and stored.

Conclusion

Congratulations! You've just mastered the art of using Windows Task Scheduler to automate the execution of Python scripts. Here's a recap of what we covered:

- **Setting up the environment**: Identified the correct Python executable path.
- **Writing a simple Python script**: Created a script to confirm that Python was running correctly via Task Scheduler.
- **Configuring Task Scheduler**: Set up a task to automatically run the Python script at a specific time.
- **Monitoring the execution**: Verified that the task ran successfully and troubleshooted any issues.

This process allows you to run Python scripts at specific times or intervals without manual intervention, which can automate data processing, reporting, or other tasks on a regular basis.

Potential Enhancements:

- **More complex schedules**: Set tasks to run daily, weekly, or at custom intervals.
- **Email notifications**: Configure Task Scheduler to send email notifications for task completion or failure.
- **Advanced triggers**: Use system events or other triggers to run tasks.

By mastering these techniques, you can streamline your workflow and eliminate repetitive tasks, ensuring that your Python scripts run automatically and consistently.

* 9 7 9 8 2 1 8 5 8 9 9 8 1 *
SD - #0066 - 140926 - C0 - 279/216/10 - PB - 9798218589981 - Matt Lamination